READING WITH JOHN CLARE

Sara Guyer and Brian McGrath, series editors

Lit Z embraces models of criticism uncontained by conventional notions of history, periodicity, and culture, and committed to the work of reading. Books in the series may seem untimely, anachronistic, or out of touch with contemporary trends because they have arrived too early or too late.

At least since Friedrich Schlegel, thinking that affirms literature's own untimeliness has been named romanticism. Recalling this history, Lit Z exemplifies the survival of romanticism as a mode of contemporary criticism, as well as forms of contemporary criticism that demonstrate the unfulfilled possibilities of romanticism. Whether or not they focus on the romantic period, books in this series epitomize romanticism as a way of thinking that compels another relation to the present.

Lit Z is the first book series to take seriously this capacious sense of romanticism. Its books explore the creative potential of reading's untimeliness and history's enigmatic force.

READING WITH JOHN CLARE

Biopoetics, Sovereignty, Romanticism

Sara Guyer

Fordham University Press
New York 2015

Fordham University Press also publishes its books in a variety of electronic formats. Some content that appears in print may not be available in electronic books.

Visit us online at www.fordhampress.com.

Library of Congress Cataloging-in-Publication Data

Guyer, Sara Emilie.
Reading with John Clare : biopoetics, sovereignty, romanticism / Sara Guyer. — First edition.
pages cm. — (Lit z fup)
Includes bibliographical references and index.
ISBN 978-0-8232-6557-2 (hardback) — ISBN 978-0-8232-6558-9 (paper)
1. Aesthetics—Political aspects. 2. Biopolitics. 3. Clare, John, 1793–1864. I. Title.
BH301.P64G89 2015
821'.7—dc23

2014047801

Printed in the United States of America

17 16 15 5 4 3 2 1

First edition

For Sadie and Solomon

Contents

Acknowledgments

This book originated in and drew sustenance from both institutions and individuals. While the work of writing happens outside of lectures, department meetings, classrooms, family vacations, trips to the gym, and dinners that run late into the night, all of the things that take the place of writing seem to have nourished it (and me) in ways that are profoundly difficult to measure.

I first decided to write about John Clare when I taught a course on "Romantic Natures" as a postdoctoral fellow at the University of California, Irvine. Clare was not a poet that I had studied in the classroom, but the course—and a memorable car-ride conversation about Clare and prosopopoeia with my fellow commuter Steven Miller—left me hooked. This book would not exist if I had not had the opportunity to teach something I did not already know. At Irvine, and then at the University of Wisconsin–Madison, my students have been constantly open-minded and usually blown away by the experience of reading Clare. I want to acknowledge in particular the students in my courses on "Romanticism and the Poetics of Homelessness" and "Romantic Autobiography," where Clare held a central place.

My department chairs in Madison, Michael Bernard-Donals, Tom Schaub, Theresa Kelley, and Caroline Levine, offered warm support both personal and professional, which helped enormously in the completion of this project. The deans and associate deans of the College of Letters & Science, Gary Sandefur, Magdalena Hauner, John Karl Scholz, and Susan Zaeske, all encouraged my intellectual and institutional projects: I thank them. I am grateful to the Wisconsin Alumni Research Foundation and the Graduate School of the University of Wisconsin for investing resources into this project. Nancy and David Borghesi, friends of UW–Madison who have now become personal friends too, offered tremendous support for my research, as did the George A. and Eliza Gardner Howard Foundation, which provided me the resources to spend a year (half of it in Paris) devoted to completion of this book.

The Andrew W. Mellon Foundation granted me the opportunity to organize a John E. Sawyer Seminar on "Life in Past and Present" with Rick Keller. This might just have been another distraction from writing, but instead it turned out to generate new lines of inquiry. I am thankful to the participants in the seminar, who shared their work often at is earliest stages; to Rick, for introducing me to scholars and fields that I would not have otherwise known; and to Amanda Jo Goldstein, who spent the 2011–12 academic year in Madison as the seminar's postdoctoral fellow.

At about the time that I began writing this book in earnest, I also began to direct the Center for the Humanities at UW–Madison. That work has shaped this project profoundly. In addition to thinking constantly about audiences, disciplines, and the future of the humanities, and talking with scholars, writers, curators, and artists many of whom have now become friends, I also have had the pleasure of serving on the International Advisory Board of the Consortium of Humanities Centers and Institutes, which has been the source of so many rich conversations and experiments in thinking. In particular, I acknowledge Srinivas Aravamundan, Ian Baucom, Margaret Kelleher, Michael Bérubé, Kathy Woodward, and Debjani Ganguly. Jim Chandler, who is the other romanticist on the Board, had many thoughts about this project at the early stages, and talking with him about it and everything else has been enlightening, but also so much fun.

In Madison, Mario Ortiz-Robles and Guillermina de Ferrari are much more than colleagues; they are true friends. For insight, encouragement, and conversation, I thank Susan Friedman, Russ Castronovo, Monique Allewaert, Fréderic Neyrat, Alex Huneeus, Ion Meyn, Terry Kelley, Caroline Levine, and Rob Nixon. Collette Stewart and Jeff Liggon show me weekly that thinking and moving are intertwined. In Paris, Will Bishop made Aux Deux Amis, where we had lunch nearly every Friday during the spring of 2012, more than live up to its name.

I have learned a great deal from Brian McGrath and taken great pleasure in thinking with him about the futures of romanticism and reading. Ian Balfour, Anne-Lise François, and Marc Redfield are superb fellow travelers. Likewise, Michael Cobb, Rebecca Walkowitz, Henry Turner, Robbie Miotke, Catherine Malabou, Mike Witmore, Kellie Robertson, Jane Simon, Steven Miller, Jacques Lezra, and Susanne Wofford remind me of the diasporic condition of twenty-first-century academia. Phil Lewis's sage advice lingers. At various stages, Jack Dudley, Lenora Hanson, and Jared Seymour have provided helpful research assistance. Jon-Paul Carr, Northamptonshire Studies Manager at the Northampton Central Library, obtained for me the image of Clare's journal. I am grateful for his timely and professional assistance.

I benefited from presenting versions of every chapter to colleagues who offered warm feedback and useful criticism. In particular, I would like to thank the Department of English and the 18th and 19th Century Workshop at the University of Chicago; the English Department at Dartmouth College; the Group for the Study of the History of Ideas and the Center for the Study of Psychoanalysis and Culture at the University of Buffalo; the Humanities on the Edge Lecture Series and the Department of English at the University of Nebraska; the 18th/19th Century Colloquium and the Robert Penn Humanities Center at Vanderbilt University; and the Obermann Center for Advanced Studies and the 18th/19th Century Interdisciplinary Working Group at the University of Iowa. I thank my hosts, including Bill Brown, Pat McKee, Roland Végső, Marco Abel, Matthew Rigilano, Scott Juengel, Eric Gidal, Teresa Mangum, and their colleagues, Colleen Boggs, Tim Dean, Jay Clayton and many others, whose questions and comments contributed to my thinking. I also have presented this project at many conferences of the North American Society for the Study of Romanticism and the Modern Language Association. Ian Balfour, Jacques Khalip, Tilottama Rajan, Barbara Spackman, Alastair Hunt, Matthias Rudolf, and Jan Plug chaired these panels, and I am grateful to them and to the audiences and fellow panelists they convened for generous engagement.

Forest Pyle and David Clark read the entire manuscript and offered comments as rigorous as they were generous. Tres's support and style has been integral to my thinking for a very long time. James Griffioen agreed to allow one of his images to appear on the cover of this book—with the more or less sole provision that it not be used to disparage Detroit. I thank him for helping me to register another way of seeing home otherwise.

For their love and support, I thank my family, especially my parents, Cheryl and Dan Guyer, who have been a constant force for the good. Erica Guyer and Danny Franklin, Jonathan Guyer, Susan Waterfall, and Leila Straus were there throughout. I am sorry that Mickey Straus, who found so many ways to celebrate beauty and art, is no longer here for this and much, much else.

Just six days after she wrote to say of this manuscript, "It's ready to go, isn't it?" Helen Tartar died tragically in a car accident. I am very sad not to have had the chance to work with Helen at this book's final stages or on future projects, including the new series that this book launches. The world of thinking and reading is much emptier without her, and her death forces all of us who care about books and criticism as a means of navigating difficulty to invent ways of supporting important work that may appear untimely. I am enormously grateful to Tom Lay (and his colleagues at Fordham) for taking up this project with speed, interest, and thoughtful engagement in a time that for so many of us was one of crisis.

My own reflections on home, the American suburbs and the complex sense of belonging and displacement that thrives in John Clare's poetry, but also in diasporic communities that live in and across the country's middle zones, led me initially to think about alternative cosmopolitanisms and alternatives to cosmopolitanism. In addition to Madison, this book was written in substantial part in a place that is not my home (Paris), although where I feel at home, and during the final stages, I came to write and think in rural Wisconsin, in a setting—with a meandering creek, ample woods, and many, many birds' nests—that taught me to read the daily changes in the fields and flowers. I believe that Clare would find this place familiar. The movements between Paris, Madison, and Ridgeway, which are far from unidirectional, mark the trajectory this book follows. All along—offering laughter, wisdom, patience, and love—Scott, Sadie, and Solomon Straus have accompanied me and made life better. There is no home that I can imagine without them.

Versions of three chapters have appeared elsewhere. Sections from the Introduction and Chapter 1 appeared as "Biopoetics, or Romanticism" originally published electronically on the University of Maryland's *Romantic Circles* website in a special issue devoted to "Romanticism and Biopolitics." I am grateful to journal editors, as well as to the editors of the special issue, my former students Alastair Hunt and Matthias Rudolf, for their invitation to contribute. A version of Chapter 3 appeared as "Figuring John Clare: Romanticism, Editing, and the Possibility of Justice" in *Studies in Romanticism* 51, no. 1 (Spring 2012): 3–24. I would like to acknowledge the Trustees of Boston University for permission to reproduce this essay. The image in Chapter 2 appears here thanks to the John Clare Collection, Northamptonshire Libraries and Information Service.

Clare emerges for readers in this society as a displaced, marginalized poet whose reputation is gradually being rehabilitated. . . . But it could be that Clare—shy, feral, intensely gifted—will never be redeemed from all the neglect and mutilation he has suffered.

—Tom Paulin

Introduction: The Life of Reading

In his last lecture of 1975–76, Michel Foucault focused on "power's hold over life," and in particular the emergence in the nineteenth century of sovereignty as a power over life, rather than death, sovereignty as "the right to make live and let die."[1] As Foucault explains in the *History of Sexuality*, "The old power of death that symbolized sovereign power was now carefully supplanted by the administration of bodies and the calculated management of life," two "techniques" that Foucault identifies not in philosophy but "in the form of concrete arrangements."[2] Foucault's insight has opened up the epoch of biopower, providing the terms and frames though which everything from sexuality to human rights can been understood as occurring in the aftermath of this shift in the very significance of life itself.

British (and French) poets writing at more or less the same time as the planners and statisticians Foucault considers—that is, from the late eighteenth to mid-nineteenth centuries—might also be understood to register a new significance of life itself. As Denise Gigante has argued in *Life: Organic Form and Romanticism*, a preoccupation with life—in her case, understood as organicism, vitality, or nature—binds the poets we typically call romantic. For Gigante, the romantics were writers (like the scientists who are their contemporaries) "committed to defining and representing the incalculable, uncontrollable—often capricious, always ebullient—power of vitality."[3] This is a power that the poets also sought to categorize, calculate, and manage, if not through new forms of record keeping and sanitation then through new uses of older tropes and figures and a new conception of the meaning and life of poetry. Taken literally, poetry can be understood as another of the "concrete arrangements" or "techniques" of power for the management of life, another site of the power over life, like vaccination or the variety of emergent forms of public health to which he alludes. This is true in both a thematic and a strategic sense: Literature of the period takes the power over life as a theme, but it also takes life as its object. We would have to look no further than a novel like Mary Shelley's 1818 *Frankenstein* to find a clear example in which all of these senses of life and power appear. There we find that race, the question of the human species, education, the threat to populations, and the emergence of new projects in biomedical engineering are construed as analogous to literature itself; and literature (formed as a novel, but figured as poetry) operates as a kind of life. Indeed, Shelley famously refers to her literary fiction as a newly formed life, more or less substitutable with and even allegorized by the life form whose existence the novel traces. Thus, *Frankenstein* could be read as a vivid example of the analogy between what Foucault calls *biopower* and literary power—or literature as a form of biopower. Yet, that what is at stake in this convergence is a monstrous formation whose full frontal force never can be grasped suggests, at least allegorically speaking, the simultaneously sublime and catastrophic impact of this new power over life on both poetry and politics.[4]

We can understand this convergence—the more or less simultaneous emergence of life as the medium of political *and* poetic power, the more or less simultaneous emergence of biopower *and* romantic poetry— as an historical or terminological *accident*, rather than a series of effects with a shared cause. However, the romantic preoccupation with sovereignty (as lyric subjectivity) and poetic power as a vital force sits uneasily with Foucault's account of biopower, while sharing its constitutive terms. The lyric subject, at least as it is conventionally characterized, is a resoundingly individual formation, whereas biopower, in Foucault's account, is administrative and neither oriented toward nor executed by the individual. Moreover, recent analyses of biopower, by Lauren Berlant and Eric Santner, also have noted that critics tend to confuse distinct conceptions of sovereignty (personal, political, and theological), a tendency that the correlation between romanticism and biopolitics could even be said to repeat.[5]

I think we can go further and consider the conditions and frames that allow life at this moment to emerge as an object (both an aim and a concern) of poetry and politics, of lyric subjectivity and political sovereignty. We can begin by noticing the temporality of this emergence itself, what appears as a nineteenth-century formation, in which *making life* is the spirit of the age, at the same time may be a late twentieth-century invention articulated in and through a reading of the nineteenth century texts and contexts that have been called our contemporaries.[6] This "new" preoccupation with the power over life cannot then be said simply to occur in the nineteenth century as an index of modernity's arrival in poetry or politics. Rather, it is a retroactive formation that surprisingly conjoins two competing theoretical gestures belonging to the 1970s both of which figure a past that they recast even as they are conditioned by it.

It is the very shape of this temporal knot that led me, elsewhere, to conceive of romanticism as a *poetics of survival*, that is, as preoccupied with and producing a condition of living on, while at the same time figuring and instantiating life as beyond or in excess of the opposition between life and death. It led me to call for an alternative understanding of romanticism, one in which poetry and politics were not redemptive possibilities, but quiet acts of acknowledgment. Whereas in that earlier work I focused on the uncomfortable repetitions between romantic rhetoric and post-Holocaust testimony, suggesting both that the romantic subject already is construed as a survivor and that lyric figures are a condition of the testimonial acts that they seem to undermine, this book develops my earlier thinking in a somewhat different direction by focusing on the convergence of sociopolitical and rhetorical-lyrical preoccupations with making live.

Taking seriously the shifting conception of life as the object of politics and poetics attributed to the nineteenth century and formulated in the twentieth century in Foucault's "Society Must be Defended" lectures and the first volume of his *The History of Sexuality* (1976) and Paul de Man's essays "Autobiography as De-Facement" and "Shelley Disfigured" (1979)—I consider this preoccupation with life as an instance (and instant) of contemporary romanticism.[7] This shared preoccupation with life registers the correspondence between biopolitical theory and romantic reading. The conjunction of poetry and politics becomes an occasion to revisit conventional critical accounts of romanticism and of literary theory. It leads me to return to rhetorical (or romantic) reading and to consider its relation to politics and biopolitics. This turn builds upon work by Barbara Johnson, who I consider at length in the first chapter, and Lee Edelman, who imagines another—or no—future for rhetorical reading when he describes the urgency of a repetitive rather than reproductive futurism, and in doing so both implicitly sheds light on the shape of romantic poetics and anticipates political movements that resist a programmatic approach to politics, like Occupy Wall Street. It also leads me to return to the romantic reading that contemporary critics have largely abandoned and to consider something about life that the various demographers and managers who appear in Foucault's texts (and in his only cursorily formulated and digressive account of biopower) may not be in a position to perceive or comprehend, but nevertheless continue to expose, which is the relation of *poetry* to biopolitics.[8] In other words, far from exhausted by Foucault's account of biopower and the theoretical accounts to which it has given rise, a lyric consideration of life, one formulated in and through romanticism trains us to see beyond the management of species and populations and to recognize the excesses that biopower and its institutions inherently fail to contain.[9]

While it might appear from my opening observation that I am inviting us to think about modern poets as managers, belonging to the same category as statisticians and public health officials, and that lyric sovereignty, insofar as it is focused upon making live is a mode of administration, my examples and readings will suggest instead the undoing of individual formations by the very gesture that appears to contain them. These lyric examples show that life inevitably interrupts power, and that its containment fails to sustain the newly formulated opposition between life and death that is at the heart of the shift that Foucault so compellingly describes.[10] But it also shows us that Paul de Man's all too famous understanding of "death as a displaced name for a linguistic predicament" has not yet had its due.

In what follows, I focus on John Clare (1793–1864), a poet who can be understood to exemplify a certain version of romanticism, in particular the authentic poetry and natural genius that Wordsworth and Coleridge advocate (and undertake to emulate) in the *Lyrical Ballads* and the valorization of romantic genius and creativity as madness that continue to linger in our own moment.[11] At the same time, Clare is a witness to the early operations of biopower on persons and poetry: first, through the 1809 Enclosure Acts that divided his parish in the service of better surveillance, management, and production, and later as a twenty-year inmate of the nineteenth century's emergent system of psychiatric medicine. While Clare's use of the vernacular and his display of an extraordinary, even hyperbolic localism reflect his poetry's rootedness, leading to his relative fame as a self-declared "peasant poet," he also experienced himself as a displaced person, or in his words "homeless at home." Clare's experience of the convergence of aesthetics and politics also is manifest in the strange editorial history of his poetry, which from the point of composition in the 1820s through to our own time has been subject to heavy (and sometimes heavy-handed) management by those who, at least according to the most generous reading of their operations, have aimed to amplify his voice by alternatively sustaining and effacing its vernacular mode.

I consider five overlapping concepts that could be understood to mark the contours of romanticism: the lyric subject, creative genius, organic nature, self-identity, and home. I challenge our understanding of what has been called "romantic ideology," and, more concretely, my readings of Clare—*heimliche* and *unheimliche*—track a romanticism dissociated from the violent nationalisms *and* redemptive environmentalisms of the twentieth and twenty-first centuries. These are movements in which concepts of home and belonging, or what Giorgio Agamben has called "Language and People," attributed to romanticism have played a part. Instead, I offer an alternative (even a romantic) reading of many of these foundational concepts, above all those to which we continue to hold fast as we undertake to give voice to the voiceless, query identity claims, and advocate simultaneously for cosmopolitan futures and local attachments. At the same time, I show that contemporary critical theories of biopolitics, despite repeatedly dismissing aesthetic or poetic dimensions of power as culpable ideologies, emerge alongside, and more importantly *within* the same rhetorical tradition as the romanticism that they denounce.

Giorgio Agamben reflects this position when in a short essay on "Language and People" he blames romantic ideology for allowing us to believe that we know what a people or a language might be, because this ideology defined the meaning of language in terms of people and that of people in terms of language. Thus, in Agamben's account, romanticism laid a fictional groundwork for the existence of political culture. As he explains, following a mode of explication that might exemplify what Nietzsche called *anthropomorphism*, referring to a self-affirming tautology in the absence of knowledge, "Romantic ideology . . . tried to clarify something that was already obscure (the concept of people) with the help of something even more obscure (the concept of language). Thanks to the symbiotic correspondence thus instituted, two contingent and indefinite cultural entities transform themselves into almost natural organisms endowed with their own necessary laws and characteristics."[12] When Agamben here accounts for romanticism as a theory of association that gives the appearance of symbolic organicism and internal necessity and attributes contemporary political culture in all of its violence to a set of assumptions that pervade therein, he forgets that structures of nonidentity and displacement are also at the core of romantic thinking about language and people, structures that inform—as my reading with Clare will show—the shape of its poetry. Yet, Agamben hesitates here, suggesting that this is not a fully successful operation, but rather one that hinges on the "almost." This ripple in his account demonstrates that the transformation from the obscure to the natural is incomplete. It is within the untold story of an unfulfilled romanticism that my reading of Clare opens.

Clare is important to this rethinking of romanticism and biopolitics in the first place because of his exemplarity, which I understand in the sense that Jacques Derrida and others trained us to see it, as that which is both part and not a part of (or *a part of* and *apart from*) the case to which it refers. Clare allows for an angular thinking of both romanticism and biopolitics. Moreover, the historic paucity of strong theoretically informed readings of Clare coupled with the disagreements about how his poetry should appear (which I discuss in Chapter 3), has allowed us to go on thinking for far too long about fundamental matters—the relation of nature to nurture, of who speaks and under what conditions, of what it means to be not at home within one's home—without him. While there have been several biographies of Clare, from Frederick Martin's first and now largely discredited effort, which appeared during Clare's lifetime, to Jonathan Bate's twenty-first century accomplishment, which has renewed interest in Clare among both scholars and the general public, and while excellent critics have written about Clare, John Barrell, above all, with his major contribution to thinking about the poetics of place and local attachments, there has not been nearly enough reading *of* or even, as I propose here, *with* John Clare.

Occasionally, Clare has been the object of cultural criticism, most famously perhaps in Raymond Williams's account of "green language" in *The Country and the City*. Several poets have paid him tribute, including John Ashbery, who in *Other Traditions* devotes a lecture to Clare (as the poet to whom he turns when he needs to get beyond a blockage), and Seamus Heaney, who gave a bicentenary lecture on him. Clare also is the subject of at least two contemporary works, Adam Foulds's novel *The Quickening Maze* (a finalist for the Booker Prize) and Iain Sinclair's *Edge of the Orison*, which commemorates Sinclair's own effort to follow in Clare's footsteps, an intensely literal version of what *reading with* Clare might mean.

A few theorists, not all of them romanticists, also have turned to Clare, often in order to articulate a radical departure. For example, Timothy Morton engages Clare in his account of *Ecology Without Nature*; Sigi Jöttkandt reads Clare alongside Samuel Beckett, Alain Badiou, and Jacques Lacan in her reflections on "a phenomenology of the one"; Angus Fletcher understands Clare's mode of vision to provide the terms of *A New Theory of American Poetry*; Kate Rigby reads Clare with Heidegger; and Juliet Sycharva surprisingly considers Clare in her account of "idealism in aesthetics." Perhaps the most sustained and promising critical engagement with Clare, albeit one that is somewhat quieter than these others, is that of Adam Phillips and Hugh Haughton in their introduction to the collection *John Clare in Context*, in which Heaney's lecture also appears. Phillips and Haughton frame many of the issues that will prove central to my own reading of Clare, including exile, romanticism, and radical poetry. At the same time, they point to and initiate a possibility of reading that is not focused on recovery or activist translations (red or green are the two usual modes, often in combination with one another), even if their essay, which still is only an introduction, cannot perform the reading that we still need.

This reading—and the reading that I offer here—is unconventional, neither particularly comprehensive nor typically historical. Rather, it is close, episodic, and speculative. It is lyrical in the sense of interrupted rather than recuperative. While this is a book about Clare, then, it is equally a book about the urgency and possibility of reading, about the untimely interventions that reading affords. Reading that might be called romantic, rhetorical, literary, or close has lately been abandoned in favor of counting at a distance or gliding on the surface. But, in turning away from the parasitic and absorptive modes of reading that I will both advocate and undertake here, we give up on the possibility that the field of questions to which I point above—nature and nurture, home and exile—might be thought anew and that we might undertake this thinking by using the very strategies that less imaginative readers have dismissed (or hailed) as turning everything into a linguistic predicament.[13]

While I will not go so far as to say that this book is a manifesto for close reading today, it does offer a response to those who have too quickly abandoned the difficulty and discomfort of close reading, those who have dismissed it as expired, extinct, and belonging to another era—as dead rather than living on. I already have said that my approach is antiredemptive, and so I should be clear that this is not an attempt to bring back to life that which has been given up for dead, even if such efforts at mere acknowledgment may turn out to be more challenging than simply saying this is so. Rather, it is an effort to stay with, to recognize, and even to accentuate, when needed, what remains, what—to use an overdetermined metaphor here—survives of reading. This is part of what is at stake in biopoetical thinking.

In the first place, this book exposes a key, unaccounted dimension of the theory of biopolitics, one that compels a rethinking of its critique of poetry and an attendant reconsideration of romanticism and its concepts. If articulated as a formula, it would say that *biopolitics is a biopoetics*. I also offer a second meaning of this book's subtitle, which is that biopoetics is a possibility of rhetorical (or romantic) reading. Sometimes dismissed as immune to urgent contexts, indifferent to historical archives, or merely self-referential, rhetorical reading nevertheless demands the sort of discomfort, untimeliness, and taste for difficulty that remains urgent when the world around us has become ever more adept at counting, programming, and graphing and in doing so has lost a taste for the apparently quaint, perhaps human all too human, act of reading and the uncertainties in which it forces us to live.

Clare inhabits a compelling place here in part because few have spent much time reading him at all. And there is so much—really too much—to read. In this sense his corpus, and its multiple variations, which includes poems that have been too heavily edited and "cleaned up" as well as those that have been rigorously unedited, recovered in their rough state absent punctuation and grammar and riddled with misspellings and errors, is one that seems well suited for reading at a distance. In fact, despite the general paucity of criticism on Clare, there have been at least two efforts at quantitative analysis of work, and more are underway.[14] The earliest of these is Barbara Strang's 1982 study "John Clare's Language," which appears as an appendix to R. K. R Thornton's 1982 edition of Clare's *The Rural Muse* (1835). In her essay, Strang undertakes to identify what is specific to Clare's language and ahead of the high-throughput methods that have become possible only with advanced computation, she uses quantitative methods drawn from linguistics, namely frequency studies, calculations of the rate at which words, or classes of words recur, in order to develop her analysis. Strang's essay is at its most exciting and provocative when it opens the door to new forms of close reading. So, for example, Strang observes that across his poetry Clare regularly uses the possessive genitive (or *–s* genitive) in an unconventional way, allowing inanimate objects or nouns to inhabit a position typically reserved for persons. Strang here is identifying a linguistic anomaly that may—or may not—be a personification. Her observation, which is part of a far-reaching consideration of the distinctness of Clare's language, of whether it bears a signature, and what its signature might be, remains within the realm of a proto-statistical study of frequency. But does this counting tell us that Clare personifies or that he does not? Does it explain why, as Strang claims, his readers found his use of *–s* genitives with inanimate and abstract nouns "objectionable"? Moreover, if it does not do this, could it? And what conception of the difference between the inanimate and the animate, the abstract and the concrete does it bear?

More recent efforts at quantitative reading have emerged in response to a tradition of close reading. Yet the formulation of this binarism is ironic enough to cast doubt on the seriousness of the opposition. In contrast to readings that make sweeping claims on the basis of single texts, small excerpts, and isolated lines (think here of the role that Wordsworth's "Boy of Winander" has had in de Man's corpus and that of his students), distant reading takes on large corpora, and as such has become the signal of rigorous, anti-ideological reading, able to cut across assumptions about authorship, periodization, genre, and taste. In other words, it is doing precisely the work that at a certain moment close reading and *theory* took as their charge. But, in returning to close reading and thus turning away from the models of reading that are swiftly populating literature departments, academic conferences, and reading lists, I want to suggest that the questions that close reading (or *just* reading) can be called upon to answer are at least for this reader the ones that most matter, and the ones that many critics have forgotten close reading can be summoned to engage. These are not questions of ubiquity or number, even if I can get excited when counting can show, for example, that we have been misreading Shakespeare's genres or that our entire conception of the rise of the novel is misguided. They are instead questions about rhetorical formation and deformation, about the relation between what Barbara Johnson called "persons and things," questions, to put it bluntly, about the incalculable rather than the calculable, about the history of the event. Now, I am willing to accept that quantitative gestures might sometimes help us specify incalculability and they can give us a taste for the mathematical sublime, but can they help us to track Clare's experience of homelessness and the perspective that this experience offers to our own? Can they accommodate productive anachronism, excess, lyricism, poetry, or an understanding of language as inherently unstable and ironic? Franco Moretti at least in some sense seems to think so.

Moretti is Clare's other quantitative reader, as he includes a brief discussion of Clare in *Graphs, Maps, Trees*. When Moretti sets out to understand what literary maps can do, he turns to Mary Mitford and to John Clare. His reading of Clare, however, is a reading of Barrell's *The Idea of Landscape and the Sense of Place 1730–1840*. Specifically, he sets out to test—by mapping—Barrell's claim that depending upon one's perspective, Clare's parish, Helpston, appeared either as a circular system (to those who lived there) or a linear one (to those passing through), and that these different geographic systems also reveal different ideological positions, a difference that maps archive. But by establishing a frame of understanding that conceives of maps alternatively as circular or linear, Moretti also misses one of Clare's most well known—and certainly most frequently discussed—accounts of his experience of Helpston, one that I will discuss at length in the final chapter.

In that episode, Clare recalls accidentally crossing the road that surrounds his village; he aims to reach the horizon, gets lost and confused, and has an uncanny experience of not recognizing his own church when he sees it from a new perspective. This disorientation shows how Clare's experience of his village fits uncomfortably into both circular and linear systems, but is instead marked by fantasy. Displacements, even minor ones, lead him to experience his home neither as a center nor as a point within a larger itinerary, but rather as a foreign land. Of course, the question Moretti is asking is not about Clare, but rather about what a map can tell us, and as distant reading goes, Moretti is acutely aware of the fact that things begin rather than end with counting, and that the critique of ideology is what counting makes possible. This might mean that the distance between Moretti and the close readers is not so great (and indeed, this is the topic of emergent research by Daniel Stout), but the rise of quantitative methods in the humanities has nevertheless gone a long way to persuade contemporary scholars that close reading is passé. In what follows, I will at times directly, at times indirectly, undertake to show that this is an error, and that untimely reading—close reading—or uncomfortable reading has still to come into its own.

1. The Viability of Poetry

On December 28, 1841, John Clare became an inhabitant of the Northampton General Lunatic Asylum, one of the genre of institutions that Michel Foucault identifies with the management and restoration of the population.[1] Clare wrote extensively while in the asylum, where he went largely unmonitored. While much of his poetry of this period (he was there until his death in 1864) has not been preserved, an undated lyric "To Mary" is one of the hundreds of poems that survive. Clare addresses the poem to a girl that he loved in childhood, but whose father intervened because he was convinced that their marriage would leave his daughter in poverty.[2] While Clare went on to marry and have children with another woman, Patty Joyce, he remained irremediably attached to Mary, as is evident in the journal that he wrote during his escape from his first asylum, which concludes with a letter to Mary Clare whom he addresses there as "My dear wife."[3] In the journal, Clare explains that he has been told of Mary's death, but that he simply does not believe it, complaining that it is just an "old story of her being dead six years ago which might be taken from a bran new old newspaper."[4] Clare remains convinced that Mary is alive and often acknowledges that he has two wives.[5]

"To Mary" opens with the chilling announcement that while the poem's subject (Clare) sleeps and wakes with Mary, she is not there.

To Mary

I sleep with thee and wake with thee
And yet thou art not there;
I fill my arms with thoughts of thee
And press the common air.
Thy eyes are gazing upon mine
When thou art out of sight;

My lips are always touching thine
At morning, noon, and night.

I think and speak of other things
To keep my mind at rest;
But still to thee my memory clings
Like love in woman's breast
I hide it from the world's wide eye
And think and speak contrary;
But soft the wind comes from the sky
And whispers tales of Mary.

The night wind whispers in my ear,
The moon shines in my face;
A burden still of chilling fear
I find in every place.
The breeze is whispering in the bush
And the dew-fall from the tree,
All sighing on and will not hush
Some pleasant tales of thee.[6]

The poem candidly acknowledges the failure of poetic apostrophe to solve the problem of absence or death ("and yet thou art not there"), and likewise, the failure of absence or death to exhaust the usefulness of direct address and the fictional presence that it remarks and effects. Mary emerges as both too resilient and utterly absent. One could say that her spectral presence is the effect of psychosis or delusion, the result of misunderstanding the difference between embodied and disembodied forms, life and language. And one could go even further to say that *this* confusion is not just a conventional outcome of lyric surmise, but rather the source of Clare's incarceration. Read in this way, "To Mary" stands as a form of autobiographical and psychological evidence, rather than a dramatization of the simultaneous power and weakness of literary figuration. And yet, at the very moment that it is taken as autobiographical evidence—as a sign of Clare's illness—it reveals the impossibility of poetry, that is, the impossibility of the forms of uncertainty that define the lyric. Clare's "To Mary" reflects the limits of rhetorical and biopolitical accounts of poetry when dissociated from one another, and it opens not only new ways of reading Clare, but more significantly, new ways of thinking about reading the rhetoric and politics of life. In other words, Clare's poem offers a radical account of the relation between lyric animation and life itself in part because it shows how poetry can simultaneously sustain an attachment—can make others live—and make the life of its first-person subject virtually unlivable. It shows, further, another version of what Paul de Man called autobiography as de-facement, not the defacement of the "I," but of poetry itself. In what follows, "a long digression" on rhetoric and biopolitics (to only slightly revise Foucault), I turn first to de Man and then to Barbara Johnson, in order to reread their accounts of lyric animation, before returning to Clare and Mary. The aim is to better describe and understand the form of lyric life.

Despite having written extensively on lyric animation, romantic lifewriting, and Percy Bysshe Shelley's *The Triumph of Life*, Paul de Man seemed to have very little interest in the question of life itself. Indeed, like death, de Man takes life as a displaced name for a linguistic predicament, that is, as a figure of figure that reflects a problem of propriety and belonging. Those texts that seem most preoccupied with life (for example "Autobiography as De-facement" and "Shelley Disfigured") are not simply "about" death or even the undecidability between life and death (recall: "one moves, without compromise, from death *or* life to life *and* death"), but about rhetoric and figural language. When in his essays of the late 1970s, later collected in *The Rhetoric of Romanticism*, de Man apparently turns away from organicist accounts of language (that is, language as a vehicle or sign of life), he does so in order to turn our attention to the ideology of rhetoric (or literature) as a restorative, indeed indissociably restorative and privative, operation. Autobiography (or lifewriting) operates through a figurative movement that "deprives and disfigures to the precise extent that it restores."[7] (Returning to the example I introduced earlier, Mary Shelley's *Frankenstein* acknowledges this movement in all of the scenes where the encounter between Victor Frankenstein and his creature involves a series of faintings and restorations.) Although, de Man focuses on William Wordsworth's *Essays upon Epitaphs* and Percy Shelley's "Triumph of Life," privation, disfiguration, and restoration in his account are not matters of life and death, an assumption that would remain within an organicist paradigm, albeit a negative one, but rather matters of cognition, apparition, and image. It is sensation, and the relation between the visible and the knowable worlds, rather than life and death, that are at the core of de Man's observations.

For all of these reasons, it might seem antithetical to turn to de Man in an effort to track the relation of literature to life and develop a theory of *biopoetics*. However, despite de Man's apparent allergy to questions of life and his indifference to biological processes or political analysis on a grand scale, his understanding of figuration has laid the groundwork for other accounts of lyric figures (apostrophe, prosopopoeia) that take place in a more explicit relation to the politics of life, for example those of Barbara Johnson. Moreover, as I will argue over the course of this book, it might be time that we allow there to be a little life in de Man's readings and that we not give up on reading in order to think about life itself.

Johnson offers one of the earliest and most engaging versions of this possibility. In her 1986 essay "Apostrophe, Animation, and Abortion," Johnson draws upon de Man's account of figure to ask whether "the very essence of a political issue—an issue like, say, abortion—hinges on the structure of figure," and she considers the centrality of personification to arguments for "the right to life."[8] In some sense, Johnson just repeats de Man's gesture, suggesting that politics—and the politics of life and death—is a name for linguistic or rhetorical predicaments. But if this is what seems to be the case, in fact something else is happening. When Johnson insists that abortion is an issue not because life is sacred or because women's rights are at stake, but because it is structured like a figure, she reminds us that as soon as we think about politics in terms of rhetorical force, we also are thinking about poetry. More than this, she compels those who mistake de Man (or later Judith Butler) as claiming that it is all just language, to recognize the very power of rhetorical language in relation to life and death.[9] Johnson goes on to ask if there is "any *inherent* connection between figurative language and questions of life and death, of who will wield and who will receive violence in a given human society," which is to say, whether any use or consideration of figure is already a matter of life and death, and specifically, of the power over life and death.[10] Put in other words, she is asking whether the rhetoric of tropes and figures is indissociable from biopower.

When Johnson's essay appeared in the mid-1980s, Foucault's various discussions of "making live" in the nineteenth century remained overshadowed by his much more substantial considerations of governmentality, the body, and its discipline. Reading her essay today—after Foucault and Giorgio Agamben and their readers—the contribution is clear: Johnson registers first as a question and then as a claim that the question of life is an inherently poetic and political one. In other words, her reflection on apostrophe, animation, and abortion is an argument for the essentially poetic structure of what we later have come to call biopolitics. Insofar as politics is a matter of power, and power is a matter of violence, the essence of any political issue is the question of life and death, of *the power over life and death*, and this is where poetry (or figural language) comes into the picture. For politics in Johnson's account "hinges on the structure of figure."[11] It hangs on, depends upon, and is suspended by a rhetorical device that *makes* "present, animate, and anthropomorphic," a device that, recalling Foucault's account of political power, makes live.[12] In other words, Johnson shows that biopolitics is a figural predicament, and at the same time that apostrophe is a biopolitical predicament. Put another way, what de Man registered as rhetorical is already, *inherently*, political and, as Johnson shows *avant la lettre* biopolitical.

While this poetic conception of life has a particularly compelling presence in the late eighteenth- and early nineteenth-century lyric, and the various effusions that run throughout the poetry of Wordsworth, Shelley, and Keats among others, Johnson argues that it has an afterlife in a domain that typically appears more political than it does literary, and she allows us to see (through a reading of Charles Baudelaire) that the poetics of animation also is indissociable from politics. For Johnson, the most apropos, if uncomfortable, example of the convergence of politics and poetics is abortion itself, as evident in the rhetoric around the fetus and the fraught, rhetorically manifest relation between mothers and potential children that occurs in abortion (and its aftermath). This is a convergence that recurs in the lyric, whether or not it takes abortion as a theme. In both cases (abortion and lyric) the question of viability—of whether and at what point someone or something may be said to be able to survive—is transformed not only by new and still emergent technologies, but also by linguistic acts and literary figures that condition our conception of the ends of life. As anyone even mildly familiar with the rhetoric around abortion in the United States knows, one side builds their position upon the assumption that a fetus feels, knows, and speaks (even silently screams) like the living, that it is a form of rights-bearing life, a person rather than a mute entity whose viability is fundamentally in question.[13] Here, politics relies upon tropes and figures when life is at stake. More than this, as soon as the relationship between life and death becomes a rhetorical one, viability, which the rhetoric of personhood and life seek to settle, becomes a question, which is to say the political question of life reappears.

Johnson begins her essay by showing in a rather traditional sense how lyric apostrophes, as acts of animation that assume the difference between the living and the dead, turn out to undo the very distinctions upon which they seem to rely. Her initial examples, drawn from Baudelaire and Percy Bysshe Shelley, relay scenes in which a lyric subject addresses an inanimate object in order to endow it with the power that will retroactively animate the very voice responsible for the address in the first place. Johnson teasingly reads these canonical poems in the romantic lyrical tradition in which a male poet undertakes to obtain a voice from the outside together with poems by women in which "the question of animation and anthropomorphism is . . . given a new and disturbing twist," poems that "textually place aborted children in the spot formerly occupied by all the dead, inanimate, or absent entities previously addressed by the lyric."[14] Like Foucault, Johnson seems to register a shift in modernity's relation to sovereignty, showing that an emergent structure of animation in the nineteenth century remains at the core of political thinking in the late twentieth century; and, like Foucault, again, she is interested in rereading and recasting an earlier emergence (which we could in both cases call *biopower*) from the perspective of its violent future.[15] It is in these poems that a new, specifically biomedical uncertainty about the nature and meaning of the living and the dead emerges. It is also here that Johnson explicitly registers a continuity—or analogy—between poetry (the rhetoric of animation) and politics as biopolitics (abortion). And, at the same time, Johnson finds in these poems in which poetry and biopolitics converge the key to redescribing the past and future possibility of lyric subjectivity itself.

In the essay, Johnson focuses on Gwendolyn Brooks's "The Mother," a poem that traces the initial disappearance and belated reappearance of the first-person subject, rendering her an *object* of the abortion ("Abortions never let *you* forget"); the *addressee* of aborted fetuses ("I have heard in the voices of the wind the voices of my dim killed children"); and finally the *subject* addressing them, albeit through a citation ("I have said, Sweets, if I sinned . . ."). Even as a subject, she remains uncertain about the status of the abortion ("oh what should I say, how is the truth to be said?") and thus about her own status, insofar as she is the object of the aborted fetus's address. Brooks's poem reflects a particularly complex case, for it archives a relation in which she did not want to bring the objects of her address to life even as this animation of their fictional voices is an effect of that decision against their life. It is a decision that keeps her hearing the voices of the dead. In her reading of this exchange, Johnson argues that "the poem can no more distinguish between 'I' and 'you' than it can come up with a proper definition of life."[16] This ambivalence about the proper meaning of life, and not only about the place of the subject, is fundamental to the structure of apostrophe, and in this sense it also reveals what is at stake in thinking abortion itself: life as separate but parallel to the subject. Whatever our politics or rhetoric, from the perspective of politics or rhetoric, it is not clear whether a life that is not viable can be considered a life at all, just as it is not always clear whether a life, even if it is not viable, is anything other than a life.

From a biopolitical perspective, like the one that Agamben offers at the end of *Homo Sacer*, the fetus appears as one of those lives in which the political and the biological have entered into a domain of indistinction. For Agamben, this is only a particularly vivid example of a quotidian situation. The same could be said for Johnson insofar as this indistinction is what she identifies as the fundamental structure of the lyric, a structure upon which "the essence of politics hinges." Johnson shows that what is at stake is not only the indistinction between political life and biological life, but also that this life is indissociable from poetical life; it is in other words what de Man called a "a displaced name for a linguistic predicament."

Johnson raises some highly specific questions about the relation of poetry to life and death, the power of poetry (or language) *over* life and death, when she understands the poetic, political, and biological questions of life as thoroughly indissociable. While Johnson considers the ways that the lyric, as an exemplary form of language, can repeat and reenact the violence it aims to overcome, she also challenges conventional accounts of lyric subjectivity as they relate to life itself. She not only argues that debates about abortion can be seen as debates about apostrophe and the rhetoric of animation, but equally that lyric poetry can be seen as part of debates about abortion—or construed more generally—debates within the politics of life. She goes so far as to suggest (while attributing this suggestion to Brooks) that "arguments for and against abortion are structured through and through by the rhetorical limits and possibilities of something akin to apostrophe." And further, that: "The fact that apostrophe allows one to animate the inanimate, the dead, or the absent, implies that whenever a being is apostrophized, it is thereby automatically animated, anthropomorphized, 'person-ified.'"[17] Viability is thus a matter not only of biology, but also of politics and rhetoric. To say that life is a "displaced name for a linguistic predicament" is therefore to recognize that it is a matter of politics, that it is a matter of power and contestation over viability—over the meaning and right to life. Johnson's first insight is into the fundamentally political nature of de Man's apparent dismissal of death (and implicitly life). Moreover, in a reading that demonstrates the poetic structure of biopolitical thinking, she also shows that the lyric and its figures are the nexus of contemporary biopolitics, and that insofar as theorists of biopolitics remain concerned with questions of persons and politics they are also engaging the question of apostrophe itself.

I opened with the observation that the romantic lyric and biopower emerge together as two forms of a power over life. Reading Johnson, I also considered how politics—and biopolitics—can be seen to "hinge" on the structure of a poetic figure, and further that to recognize this structure is to see poetry and politics as two forms engaged with questions of viability. Johnson's essay does not stop here, for it challenges us to consider what happens when politics and poetry depend upon apostrophe and when making live becomes the operative mode of poetry and politics both. Johnson not only perceives the conjunction of poetics and politics but also exposes the implications of this conjunction on conventional accounts of romantic subjectivity. She reminds us that Baudelaire opens *Les fleurs du mal* by representing himself as a failed abortion and an originary survivor. She invites us to see this neither as a hyperbolic drama nor an anomaly, but rather as exposing the structure of the lyric (and political) subject. Far from a sovereign capable of action, the lyric subject is structured as a fetus, a life whose viability is permanently in question.[18] "Fetality," rather than sovereignty, is the position of the romantic lyric subject. Thus, just as Agamben considers the paradox of political sovereignty as conjoining the sovereign and the exile in the same position of exceptionality before the law, Johnson, reading Baudelaire and Shelley after Brooks and Adrienne Rich, articulates a new paradox of lyric subjectivity (and implicitly political sovereignty): the aborted fetus and the (masculine) subject inhabit the same position.

Yet, the structural similarity between lyric and political accounts of sovereignty does not account for a suggestive difference. Penelope Deutscher has argued that fetal life is not identical with the forms of "bare life" that Agamben associates with biopolitics. For Deutscher, the difference between bare life and fetal life is both temporal and categorical. The fetus produces a scene of contested life (or what I have been calling a question of viability) that emerges *prior* to loss or privation and prior to political subjectivity, if not prior to the politics of subjection. As Deutscher writes, recalling Agamben's various figures of bare life: "A consideration of fetal life does not fit the series [*Muselmann*, overcoma, etc.], as it usually is not situated at the threshold of depoliticization or dehumanization of previously politicized or humanized life. The fetus represents the zone of contested and intensified political stakes around the threshold between what some would consider 'prelife' and what is to be identified as nascent human life, meaningful human life, and/or rights-bearing life."[19] However, when Johnson reads Baudelaire and Brooks, she identifies the lyric as effecting a zone of fundamentally contested viability. Her insight into the shape of the romantic subject does not merely anticipate Agamben's figures of biopolitics, but it also already moves beyond them.

She also aims to move beyond the limits of a lyric structure that produces subjects, whether masculine or feminine, as fetuses. Drawing upon Brooks's poem as well as other poems of the same subgenre by Adrienne Rich, Anne Sexton, and Lucille Clifton, poems in which the subject is figured as a mother who is both addressed by and addresses herself to her dead offspring, Johnson suggests that if the lyric assumes a subject who inhabits the position of the fetus, then it both affirms and effaces the position of mothers. The entire history of the lyric—of poetry and politics, insofar as they are bound up with the rhetoric of calling—can be understood neither as narrative nor as genealogy, but as the repetition of an originary (or primal) apostrophe, an address to one's mother. Johnson suggests that were we to recast the prevailing accounts of subjectivity and recognize the lyric subject not as the effect of an infinite substitution ("men have in a sense always had no choice but to substitute something for the literal process of birth"), a specular trope driven by absence and loss, but rather of unsubstitutability, something like an anthropomorphism, which de Man understands as recurring to a name, we might find a path outside of fetality, permanent childhood, and masculine subjectivity.[20] Turning to the mother in order to recover her absence and reorient the history of lyric and sexual politics is one way of thinking about "making live." Recalling Deutscher again, we can see another aspect of the difference between fetality and the forms of precariousness that have come to dominate other accounts of biopolitics. Fetality reveals an enduring mode of sovereignty that operates in strange parallel and uncomfortable debt to motherhood. In this frame, not only is the relation of apostrophe to animation one of motherhood, but biopower understood as "making live" also operates in a way that relates sovereignty to motherhood.[21] Johnson allows us to see that in the nineteenth century, political sovereignty becomes bound up with maternity as the most familiar example of "making live." She reminds us that this sovereignty or position of subjectivity, far from univocal has other manifestations, including that of the mother who does not "make live" or who makes live (and hence becomes a mother) only when she animates that which she also has let die, and in some accounts even killed.[22]

On Johnson's reading, the lyric is a form of life only to the extent that it simultaneously repeats and effaces the maternal origins of life and language as address. If the lyric subject is conventionally figured as a sovereign, and if rhetorical reading exposes instead that it is variously a fetus, embryo, infant, or child, Johnson aims to break the fetus–sovereign dyad by recognizing that what has not or cannot be said is that lyric animation is akin to the very motherhood that it denies. Motherhood can be an effect, not only a source, of address; it can have an ambivalent relation to life-making; and we ought to begin to see the lyric subject not as a man or a child (which is to say a fetus), but as (or wanting to be) a mother.[23] It is as if our failure to recognize the lyric subject as mother and as a form of sovereignty obtained by life making has left us with a structure of animation that cannot but leave the subject a child (or fetus) whose viability and power is in question.

When we see the subject as a mother, because we have seen it as a fetus, we see that the subject inhabits an ambivalent relation to political sovereignty and biopower, that is, a relation bound up with "making live" and "letting die," with noncriminal forms of death, such as abortion. Johnson's radical insight into and recovery of motherhood elegantly corresponds to Foucault's account of political sovereignty as it emerges in the nineteenth century. Johnson allows us to see the mother absent (or structurally present) not only in the romantic lyric but also in the political scene in which it came to emerge. When Johnson presciently accounts for the lyric as an engagement with life and death in which politics is in question; when she recognizes the fetal structure of the lyric subject and the possibility of an effaced maternal subject, bound up in one of the major nodes of biopolitics, abortion, she also shows in a novel and anticipatory way how the lyric is a mode of biopolitics.

Whereas Johnson's reading of apostrophe focuses on the relation between political, biological, and poetic life, revealing the limits of the romantic figuration of the lyric subject as man-child, Clare's "To Mary," with which this chapter began, stages animation not from the perspective of the effaced mother, or the surviving child, but the haunted lover in whom poetry and pathology become inextricably conjoined. It is as if all of the scenes of an impossible, eternal love that the poem marshals become the image of *this* experience of poetry and its impossibility. This is not a vertical, generational fiction of specularity (whether on the model of Wordsworth's "Bless'd babe," Baudelaire's stillborn survival, or Brooks's speaking fetuses), but a horizontal one. As an apparently conventional love poem, one that assumes and reflects upon the communion of a living lover and his dead beloved, "To Mary" evokes tropes of remembrance and loss, presence and absence. As a stubborn account of survival—of life beyond life and death—it also indicates the poetry and the politics that produce and sustain a life perceived to be unfit for society. What we have here is neither the mourning—fulfilled or not—of a maternal poetics (in which we would have to include Victor Frankenstein), nor a strictly political exclusion that reveals the new politics of public health. Rather, we encounter an apostrophic poetics in which making live coincides with a refusal of loss. This refusal, this insistent animation, which poetry makes possible, and which may even be its signature effect, becomes unmanageable and emerges as a form of disturbance and pathology. It produces the subjects of biopower as poetic subjects and the poetic subject as a biopolitical subject. Here, poetry and mental illness converge through the instruments of lyric and the instruments of biopower that resemble them. In other words, the symptom and the remedy (if not the cure) are both forms of apostrophe.

If "To Mary" assumes the presence of its addressee, it also breaks the analogy between life and presence, absence and death, as well as the negative (or chiasmic) version of this analogy. Moreover, unlike Shelley or Brooks, Clare asks nothing of his addressee, he merely describes his relation to her. (It is true that the excessive nature of this description nevertheless makes it seem a creepy demand.) In a move that seems almost to concede, finally, to the absence that the act of address rejects and redescribes, a move that also evokes some of the moments of strange concession that occur in Clare's journals, the poem leaves its addressee in a position of utter nonresponsiveness.

Mary becomes the *matter* of an address rather than its addressee; Clare becomes the object of the address, not from or by her, but of her. *Mary* becomes *murmur*, and a murmur: the whispering and signing of the air: "soft the wind comes from the sky / And whispers tales of Mary." It is not Mary who speaks or whose voice he hears; rather, he hears *of* her through the whispering wind, through the rustling of the common air. This is the convention of Clare's contemporaries, Wordsworth and Shelley, but also a convention that, as Johnson shows, Brooks reworks by "textually placing aborted children in the spot formerly occupied by all the dead, inanimate, or absent entities previously addressed by the lyric." However, unlike Wordsworth or Shelley or even Brooks, Clare registers a scene in which it is *not* the addressee who speaks—directly or indirectly—through or to the wind; it is *not* Mary's voice that he hears. Rather, the wind speaks to him in *its* "own" voice *of* Mary ("The night wind whispers in my ear . . ."). This final account of an apostrophe does *not* raise significant questions about the *power* to marshal life or death (Shelley) or about the *sovereignty* of the lyric subject as poet or as mother (Brooks). Instead, it exposes the subject's impotence and haunting.

This weak or triangulated apostrophe, an apostrophe that breaks the dyadic subjectivism of the conventional lyric, also reveals another version of the poetics of life. The recurring tales that resonate in the breeze are the return of what the subject attempts to repress or conceal, figuring yet another scene of his—and the trope's—ineffectiveness. Much as he tries to say or think something else, the tale of Mary can neither be displaced, nor extinguished. While the third stanza seems to describe a scene that might be one of revelation—that might expose the ghostliness and disturbance of this relation, between a lover and mere air, as if he finally were to admit the truth of her absence and the mistake of his belief in her presence, what is revealed here is instead another form of haunting. The turn begins with the whispering wind, and by the third stanza, the world has become animated in a speaking scene that occludes the beloved. Mary's absence is replaced by the "whispering" and "sighing" of tales, what Sigi Jöttkandt calls Clare's "mary-ing" of the world.[24] The poem that opens with a lyric address by the end turns the subject into its addressee, and its initial addressee, far from disappearing, becomes only the topic of the address.

While in Shelley's "Ode to the West Wind" and Brooks's "The Mother," there is a substitution of subject and object, here we are left to ask who is addressing the subject and what has happened to the object of the lyric apostrophe? In this case we do not have a mirroring, a reciprocal animation, or even a crossing of lyric and maternal animation. Rather, when Clare thinks and speaks "of other things," when he turns from his beloved to the landscape, when he holds her in "the common air," he finds only the very resilience of this passionate attachment to nothing, absence, and impossibility—including the impossibility of poetry as other than pathology. It is a scene in which hyperanimation has riven subject and object, person and place both.

Having first explained that in wakefulness as in sleep he remains attached to one who is not there, Clare goes on to describe a relationship that endures when thought substitutes for and is more resilient than a body, and when a physical embrace of immaterial forms becomes a means of strange animation. He holds only "common air." *Common* here suggests both the quotidian and the shared; and what is both quotidian and shared—the air—is a condition of life, just as here it evokes the simultaneous endurance and absence of the living. By the poem's end, this air becomes not merely nothing or a life breath, but the place of Mary's life as a whispered tale ("But soft the wind comes from the sky / And whispers tales of Mary" or "The breeze is whispering in the bush / And the dew-fall from the tree, / All sighing on and will not hush/Some pleasant tales"). Curiously, when Clare figures this image of an embrace of absence—of no one and nothing—as an experience of clasping that which is held in common, he recalls his interest in the common in an apparently other sense: the commons as that space that belongs to no one and to all, that space devastated by the Enclosure, and transformed from a zone of shared possibility into one of private ownership and exclusion.[25] In "To Mary," the common is at once that which is necessary and that which signifies an absence; it is what remains when there is no body and no legitimate history to behold, when there are constant markers of temporality ("at morning, noon, and night") and yet no change. While the common, figured as air, belongs to no particular place, we also know that its quality and capacity for sustenance is bound up with the preservation of particular places. In "To Mary," however, this air—ultimately figured as wind and breeze—comes to trouble the dyadic relation, as if turning the embrace between lovers into another form of life.

When Clare renders himself the object of Mary's gaze, he continues to describe his relationship to Mary as one of absent presence: "Thy eyes are gazing upon mine / When though are out of sight." Yet, her gaze, far from direct or reciprocal, occurs *only* when she is "out of sight," which is to say, *always*. Absence is cast as a sign of presence. This logic of absence *as* presence reappears in the infinite kiss with which the stanza concludes: "My lips are always touching thine / At morning, noon, and night." Here the magical thinking in which absence and presence are no longer opposed is not a straightforward mode of recovery (as in Brooks), rather it approaches a violent end, for a physical relation like this one is utterly untenable. This kiss not only conjoins Mary to her lover in a permanent, atemporal relation, but also conjoins the literal and the impossible. The lyric apostrophe that produces an eternal life and bond is one that establishes a new opposition, between poetic life and its realization, which would be a death. The poem then generates a new form of relation—a new scene of making live.

"To Mary" is an example of apostrophe where the animation of the dead becomes an index not merely of imaginative power or figural possibility, but of "lunacy." Taken literally, it exposes an inherent failure to arbitrate between the living and the dead or the self and another, and this failure is both a rhetorical effect and medical symptom. Clare archives this radical privation in a poem that strangely, even perversely, registers the effects of making live. "To Mary" is not simply a personal intervention or an encounter with the personal as political, but rather a poem that bears witness to the lyric and the asylum as two scenes of managing life that both house and sustain a form of madness or haunting.[26] In fact, the relation here is even closer than first meets the eye, for this poem, like all of the poems that survive Clare's twenty-three-year internment, comes to us because of its transcription and preservation by the asylum's steward, William Knight. There is no remaining manuscript, only a transcript composed in Knight's hand. "To Mary" is therefore a poem whose survival is an effect of the very institution called upon to keep Clare alive *and* separate him from the living; it is a poem that registers not the work of "madness and civilization," but of "madness and biopower," and the apostrophic or *biopoetic* structure of both.

It is this excessive animation that leaves Clare writing from the asylum. Moreover, it is this mode of apostrophe that, like Johnson's account of apostrophe in abortion poetry, and de Man's account of death (and life) as a displaced name for a linguistic predicament leads us to rethink the relation of lyric to life, apostrophe to sovereignty, for it is anything but sovereignty that apostrophe in this instance seems to wield. The risk of this nonsovereignty is that it remains tied to institutions, like the lunatic asylum, in which the nourishment of life and violence against the living are conjoined. While the asylum protects Clare from society, preserves his poetry, transcribes, and archives it, it also keeps him from seeing his family for over twenty years (from his first commitment until his death). If this poem bears witness to a poetics of life that breaks with the dyadic model, it also can be seen to proliferate it. For when Clare describes the lyric subject as the object of address he may already, proleptically, be describing a structure of reception in which his own voice becomes a "whisper" and in which is own poem comes to us in a double form, borne by a writer who is ultimately not the poet, and another listener who may or may not be its addressee: William Knight.

Johnson's reading of Brooks after Shelley and Baudelaire allows us to witness anew the lyric structure of sovereignty—and to see in the place of the lyric subject, perceived as sovereign (Mill) or as mute (de Man), a subject whose very viability is in question. Moreover, Johnson allows us to see that to talk about life or death as a displaced name for a linguistic predicament is not to deny its force, but rather to acknowledge that a figural predicament is a political predicament, and to recognize that rhetorical questions are at the core of biopolitics, and that political questions are at the core of rhetorical reading. Clare's poem indicates another iteration of lyric subjectivity as it relates to life. "To Mary" seems to rely upon another dyad, no longer mother (living)/child (dead) or poet (living)/breeze (nonliving), but rather a more familiar apostrophic encounter between the lover and beloved.

Yet, the dyad, while utterly intense dissipates into a scene in which *nothing* and no one can be fully recovered. It reveals the pervasive uncertainty that examples of the mother and the emerging poet cannot concede. Here, uncertain viability is shared between the poet and addressee, not as one possibility of identity, but as existence itself, just as the power over life is divided between the poet, his steward, and the whispering breeze. This whispering leads not only to poetic fame or posthumous living, but also to what João Biehl has called "social abandonment" and ultimately actual incarceration. The relation of lover and beloved becomes indissociable from that of inmate and asylum, poetry and pathology. A typescript and a place in the archive replace the absent manuscript; minimal mobility and freedom from total impoverishment substitute for the loss of family; a permanent embrace replaces one impossibility for another. If these substitutions are forms of "making live," they also remind us that the rhetorical figure of this animation is apostrophe. Apostrophe, rather than remedy a pathology, instead sustains it, and becomes its form.

De Man registered the correlation between privation and animation in the lyric, and Johnson recognized this correlation at once to evoke and efface the role of the mother, rendering the fetus or infant in the position of lyric subjectivity. I want to affirm that each of these moments, which rely upon a rhetoric of animation, are not merely poetic. Rather, they indicate a condition in which questions of sovereignty and representation are simultaneously, but impossibly poetic *and* political. And further, this convergence of poetic and political questions around apostrophe, the convergence of biopolitics and biopoetics, reveals a poetry that is at once ineradicable and impossible, a poetry that is given over to murmurs, and for which murmurs reveal not only the liberating absence of a subject but also the symptom that will warrant his incarceration, that will, like the address "To Mary," masquerade as an act of animation, of "making live."

2. The Origins and Ends of Poetic Genius

Throughout his life, readers considered John Clare a genius and in doing so established an understanding of poetry and of the poet (like that of human rights and the rights-bearing man) that originates in mere life. For example, in January 1822, an unsigned review of John Clare's *A Village Minstrel*, possibly by Josiah Conder, opened with a description of Clare that turned out to anticipate his epitaph. The review begins: "It still holds true as ever, that a poet must be born a poet, he cannot grow into one."[1] The review set out to explain that the organic ideology it advocates is not, however, a claim about talent (which, he holds, *can* be acquired) or ability (which can be developed), but rather about *character*, which is presumed to be innate. Character cannot be formed by an education, upbringing, or formula, that is, by the abundant strategies of formation that circulate, as the author continues, in the emergent "age of mechanism."[2]

This original character—or genius—leads to "genuine" poetry rather than merely learned forms that are beautifully calibrated, like a steam engine or Dutch tulip.[3] The reviewer concludes: "Such a poet as John Clare, education could not have made, nor could adversity destroy."[4] In other words, not only is an innate character and aptitude for poetry without a source in experience, so too is it without an end. It is impervious to the environment, spawned by nature, rather than nurture. Clare's editors sustained this image of him as a natural, and in some respects Clare himself fostered it through poems like "Pastoral Poesy" ("True poesy is not in words") or "The Progress of Rhyme" ("'Twas like a parent's first regard / And love when beauty's voice was heard"), a vision of poetry originating with the birth of the poet and as a matter of life itself. Given the pervasive sense of Clare as a poetic genius, it is not especially surprising that the inscription on Clare's grave reads, "A poet is born not made." Yet, what is surprising is that, at least in 1824, Clare wrote that he wanted his grave to reflect a rather different understanding of the relation between poetry and life. Reading together the sketches and descriptions of the grave as Clare imagined it and Clare's actual grave, there emerge a set of uncomfortable contradictions about the relation of poetry and poetic identity to life and survival. Moreover, reading together Clare's two graves with the coalescence of birth and recognition in a political sphere reveals the poetic dimension of biopolitics—or biopoetics.

Two years after the *Eclectic Review* described Clare as the kind of poet who only could be born, not grown, and a full forty years before his actual death, John Clare, suffering from a severe, but not unusual, bout of illness and three nights of dreaming that he had gone to hell, decided to compose his will and sketch his grave. By this time, Clare had been recognized in and beyond literary circles as a significant figure, albeit a "peasant poet" who exemplified the vision of poetry that William Wordsworth advocated in his manifesto-like preface to the *Lyrical Ballads*. Clare was exceedingly poor, self-taught, if well read, and was known for his intense, spontaneous, unaffected descriptions of his parish and its environs. His poems often included terms from his local dialect and tended to avoid (or only arbitrarily use) punctuation and conventional grammar. Clare's *Poems Descriptive of Rural Life and Scenery*, where the author is identified as "A Northamptonshire Peasant," appeared in January 1820. That same month the *London Magazine* introduced Clare as "an Agricultural Labourer and Poet" and *the* undiscovered genius of the moment. Already from the title page of his first publication, it is clear that Clare emerged as a poet whose biography—life circumstance and social position—is an essential aspect of his reception, and whose condition as a poet depends upon its perceived origination in and indissociability from living. The year 1821 brought the publication of his second volume of poems, *The Village Minstrel and Other Poems*. Clare spent most of that summer in London, where he had gone to recover from another bout of severe illness, apparently brought on by malnutrition and exhaustion. This was only his third visit to London, and he arrived melancholic and confused. His doctor prescribed quiet and removal from society, but while he was there, Clare nevertheless managed to meet Charles Lamb, Samuel Taylor Coleridge, Thomas de Quincey, and William Hazlitt, along with his publishers, benefactors, and friends. Clare also had the uncanny experience of accidentally colliding into Lord Byron's funeral procession on the streets, a dazzling event and a strangely overdetermined one, for years later he took it upon himself to write new installments of some of Byron's best-known poems, including "Childe Harold" and "Don Juan." By the time of this visit, however, popular interest in Clare's style of poetry ("genuine poetry") had somewhat waned. Indeed, Dr. Darling, who had cared for Clare in London, went so far as to urge Clare to give up on a career in poetry and return to his life as a laborer in the fields. It was presumably in London that Clare received the journal in which he began to write soon after his return home and in which he describes his grave—forty years before his death.

Clare conceived of the journal as a place to record "opinion[s] of things I may read or see & set down any thoughts that may arise either in my reading at home or my musings in the Fields."[5] Both Margaret Grainger and Jonathan Bate have described the physical form of the journal. It was a student journal, published—and likely given to Clare by—John Taylor and James Hessey, Clare's own publishers. The book had prescribed sections for each date, a strong moral inscription about the virtues of journal keeping, and a selection of model extracts from Gibbon.[6] The sections are undated so that the writer could begin to keep a journal at any point, but they are small. As Grainger notes, each section (apart from that for Sundays) was only four and a half inches by one and seven-tenths of an inch, which might explain why Clare outlines the details of his desired burial and his will in the section designated as an appendix, rather than the main body of his text. Clare begins to write in his journal on September 6, 1824, and toward the end of September, again suffering from illness, he worries about "the anguish of leaving my children & the dark porch of eternity" and the possibility that his children, also ill, might leave him, writing that it "makes me disconsolate & yet how happy must be the death of a child."[7] In early October he worries constantly about the language of his will, and then, finally, decides to write it all out on October 8:

> Very ill to day & very unhappy my three Children are all unwell had a dismal dream of being in hell this is the third time I have had such a dream—as I am more & more convinced that I cannot recover I will make a memorandum of my temporary conserns for next to the Spirtual they ought to come & be attented too for the sake of those left behind I will insert them in No 5 of the Appendix.[8]

Just a week earlier, Clare had drawn up another will, but got hung up on the phrasing, admitting, "I don't understand the expression in it of my 'Sons & daughters & *their respective Representatives*' & shall have it alterd" and then a couple days later, "I have again reflected over my new will & I believe the expression of '*and their respective legal representatives*' is wrong so that I shall alter it as soon as it is returned." These concerns about managing his meager estate and about literary fame more generally are the focus of the *5th Appendix* to Clare's month-old journal.

In the appendix, Clare focuses primarily on the state of his finances—

what remains, but also what has been lost—and on his manuscripts. Yet, in the final paragraph, Clare turns from managing his estate to imagining another aspect of his afterlife: his grave. He explains:

> I wish to lye on the North side [of] the Church yard just about the middle of the ground were the Morning and Evening Sun can linger the longest on my Grave I wish to have a rough unhewn stone something in the form of a mile Stone so that the playboys may not break it in their heedless pastimes with nothing more on it than this Inscription: *Here Rest the Hopes and Ashes of John Clare*.

Clare's desires are exceptionally practical, he wants a sturdy, upright grave, nothing fancy, but also calculating: he wants a place in the sun and a marker that will endure not only the elements, but also the village boys and their antics. It is a desire that reflects an understanding of place—of the local conditions—not only in life, but also in death. In order to describe his grave, Clare both writes out the inscription and sketches an image. He draws a simple stone with shading to give depth to the flat page and covers its entire face with the inscription: "Here Rest the Hopes and Ashes of John Clare." To the left of the sketch, Clare writes in a hand that many have described as "cramped" that he would like the grave to be undated, and to the right he dates the journal entry (October 8, 1824) and adds the quotation from Ecclesiastes: "Vanity of vanity's all is Vanity."

Clare's explanation for why he wants no date is somewhat convoluted. He writes: "I desire that no date be inserted there on as I wish it to live or dye with my poems and other writings which if they have merit with posterity it will and if they have not it is not worth preserving." Clare seems to suggest that it is not biological life, the life of his body, but literary life or posterity that matters here. His posterity will depend *not* on a grave that marks his mortal existence but on a marker that replaces the mortal body with immortal, or at least timeless, verse, an inscription that recognizes the body as mere ash, and the grave not as a marker of survival, but of death.

Geoffrey Hartman and Paul de Man, following and reading William Wordsworth's first *Essay Upon Epitaphs* (published about fifteen years before this journal entry), taught us to recognize the ambivalent status of life and death that attends epitaphic writing, and de Man in particular described autobiographical and epitaphic writing as two inevitably failed attempts at writing beyond finitude. Clare's proposed epitaph, as he figures it in his journal, assumes that posterity is an effect of poetry. Poetry becomes a form of life that an epitaph *affirming* the limits of bodily life makes possible. By effacing life as the life of the body, suggesting that the body is mere, undated ash, and attaching life and afterlife to poetry, Clare—at the moment that he imagines his grave—also conceives of *poetry* (not the body or organism) as the vehicle of life. At the same time—as we will see—what defined (and continues to define) Clare's poetry and its posterity is precisely its indissociability from his bodily life. This tension between two notions of life is also a tension between two notions of meaning and work and two ways of remembering and forgetting.

Clare expresses an ambivalent relation to his own memorialization, but not a particularly remarkable one: Forget my life in order to extend it; let my poems, not my body, bear my life and legacy; substitute poetic infinity for mortal finitude. Yet, if this is what Clare seems to want when he imagines an undated grave and conceives of posterity through poetry rather than through a headstone that would bear his name and date his death, the actual syntax here is not quite clear. What Clare says will live (or die) with his poetry is *not* his name, which he wishes to appear on the grave, and which as the name of an authorship, he recognizes is no longer identical with his body, but his *date*, which he does not want there (and which, practically speaking, despite Clare's predictions is at this point still unknown): "I desire that no date be inserted there on as I wish it to live or dye with my poems and other writings." Curiously, the articulated subject of survival here is *not* Clare, nor is it the fictional animation of a speaking voice through the inscription of a legible name, but rather it is a date—the date of Clare's birth and death. The "it" here could be an accidental inscription of the sort that abounds in Clare's writing. He is admittedly ill and failing in both body and mind, and he is known for a grammar so inventive and a sentence construction so loose that this might not be what it seems. Indeed, it might not be too far-fetched to suggest that Clare's wish is more conventional, and that he really means to say: "*I* wish to live or dye with my poems and other writings." It is worth remembering that this is the author of two poems called "I Am," and on this account, we would understand Clare to say that he wishes to separate his lived existence from his literary posterity, which already would have begun as soon as he began to publish, affirming the bifurcation of textual and biological accounts of living. Yet, this neutralizing misreading falters when Clare again repeats the pronoun, explaining that if his works "have merit with posterity *it* will and if they have not *it* is not worth preserving." Again, one would expect that the referent here would be the name that would live or die, where biological or social life is traded for textual or literary life. This is precisely *not* the case, though, for it is the name, not the date, that Clare inscribes on his grave, the name that lives apart from the dates of a biological life on the stone. Put another way, it would seem like the absence of the

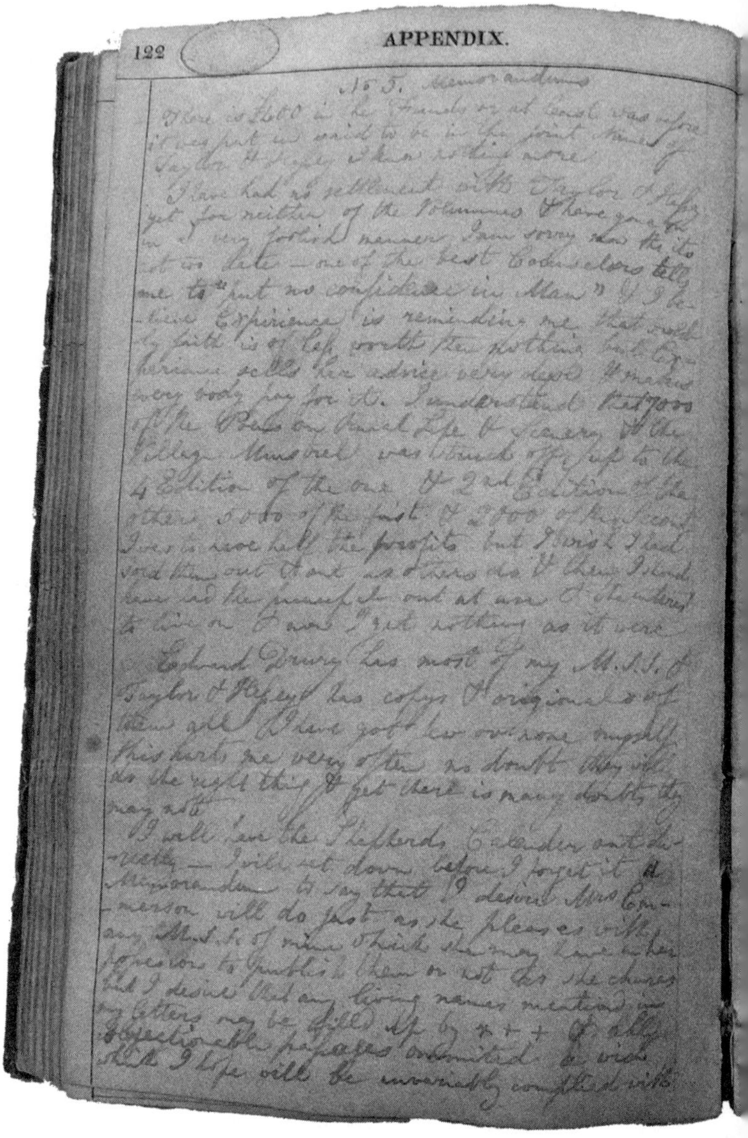

date ("it") is what *allows* the name to endure beyond dates, to live beyond life. It establishes not the conjunction of poetry and bodily life, but rather a division between the two.

By focusing on the survival of the date, not the name, and on open-ended futurity, Clare conceives of poetic life as divided from bodily life. Again, this may not seem especially original. In fact, in a compelling account of

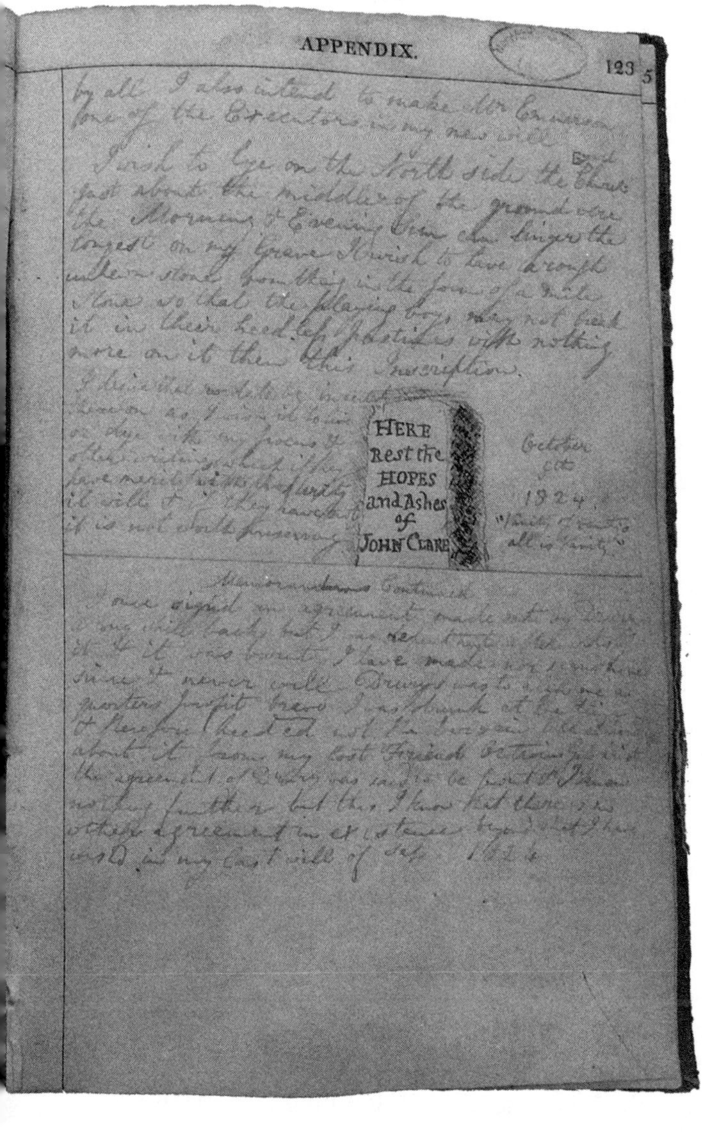

romanticism, Andrew Bennett has argued that the major poets' obsession with posterity produced their posthumous lives. Yet, for Clare there is an important difference. It is precisely the relation between his poetry and his fragile bodily life that had been the source of his fame. Clare obtained recognition, not—or not only in his name, but as the *Northamptonshire Poet*, as a peasant poet and laborer. For Clare true posterity would be freedom from the logic of genius and freedom from the name: life in which survival in poetry coincides with the forgetting of the body to which it refers.

Clare envisions the untimely survival of his dates in a text that follows the logic of quotidian dating: a journal. To recall, the description of his grave and its absent date appears to the left of the sketch, but to the right of it, Clare dates his entry and adds the line from Ecclesiastes, "Vanity of vanity's all is Vanity." The quotation here seems to affirm the vanity of focusing on one's posterity in more than one sense. It is a vain, or *self-involved* act to talk about one's fame in this manner—and it is vain, or *useless*, to make these preparations, as we will see. Further, as Clare's example suggests, it may be sheer vanity to imagine the separation of bodily life from poetic afterlife. That Clare would wish this separation is no doubt in part due to his unidealized sense of peasant life and its urgency, and perhaps equally to a complex ambivalence about the source of his fame and its contingency upon the latest thing (whether in his time or equally in our own). Yet, such an aspiration is vain not only because the reception of Clare's poetry from the 1820s until the present is indissociable from a consideration of his life, and a preoccupation with questions of organicism, nature, and local attachment, but also because at the very moment that he intends to inscribe this difference, he also undermines it.

The sketch of the tomb, the explanation that he wants no date on his grave, and the inscription of a date and closing quotation are positioned outside of the narrative of the memorandum and alongside one another. Taken together, they seem to constitute the parts of a conventional epitaph: name, date, biblical quotation, as if the dates and quotation inscribed in the journal represent the repressed tomb. Here, we have not (or not only) the realization that inscribing an end in the midst of life is as privative as it is restorative, the model of writing that in their own ways Hartman and de Man developed in the 1980s, but rather a conception of the impossibility of separating bodily life and poetic afterlife, the date and the name.

To summarize: Clare imagines a grave that would be undated, and yet he inserts, as if by accident, the date of the journal entry, an unremarkable, quotidian date, rather than the date of death, which is unremarkable in another sense. The appearance of the image of the undated tomb within the journal and the inscription of the date of the memorandum directly next to the sketch confuse dated, quotidian writing and undated, monumental writing, the temporary and the timeless, writing that belongs to a life and writing that commemorates it and its end. The journal, where the memorandum appears, is a text entirely organized around dated entries. Yet, in the convergence of these two modes, the journal also introduces the possibility that every date—every day—might bear an event and that those days and dates, taken together, might amount to a life to remember, the very life that the grave Clare envisions is constructed to forget. The seemingly accidental insertion of a date at the moment that Clare insists there should be none, at the moment when a date would attach a name to bodily life, the appearance of the grave within a journal, reflects another conception of death than the horizontal and eventful one. In his journal, but also through it, Clare reflects the attrition of life in everydayness. This is not simply melancholic or poetic suffering, but rather the accumulation of details and complaints that reflect the corrosion of life in labor, illness, displacement, and depression. These struggles with nightmares, poverty, misrecognition, and illness are the condition of bodily life, rather than its culmination, just as bodily life is the condition of this affliction. More than this, as a genre, the journal assumes that life is a series of minor, consuming events. While Clare intends to fill his journal with readings and observations, the journal also becomes a compendium of symptoms, ailments, and "newspaper odditys." Clare's undated grave, far from marking a biological death that leaves open the possibility that he will continue to live a life beyond life in poetry, *fails* to mark the irrelevance of his biological life and fails to separate a body that is left to die from a body of poetry that is made to live. The marker becomes merely an entry, albeit an appended and especially fraught one, in a journal that will continue for another year, and a life that will continue for another forty.

When Clare died in the asylum in 1864, his family was impoverished, and despite his wishes, he did not receive the grave he wanted. He is buried in Helpston, as he expected he would be, in a long, low vault near to his parents. Upon Clare's death, plans for a pauper burial were underway until friends intervened to transport his body—the body of the Northamptonshire Poet—back to Helpston. Clare's remains were buried right away in St Botolphs, albeit not quite in the spot that he once imagined, but it was only in 1867 that the grave obtained a stone. Clare's gravestone bears two separate inscriptions. On one side: "Sacred to the Memory of John Clare The Northamptonshire Peasant Poet Born July 13 1793 Died May 20 1864." And on the other "A Poet is born not made."

These two inscriptions reflect two conceptions of the relation of poetry to birth and bodily life. The first is specific to Clare's life and marked by his actual dates; the other, more general and citational, naturalizes poetic production as an effect of birth. The one figures Clare as the poet of a particular place, a particular moment, and a particular form of lived experience, and the other figures Clare as exemplifying a concept of poetry and the poet as an inherently talented being and natural genius (despite the distinction between character and talent in the *Eclectic Review*). Both statements, each in their own manner, seem to indicate Clare's position as a product of place and birth (nature rather than nurture, or to use the language of contemporary neuroscience, inheritance, rather than plasticity and epigenetics); both, in other words, figure Clare—and the poet—as a body. One associates birth with an attachment to a particular place, in this case with a natural relation not to a nation, but to a region, its knowledges, practices, habits, and language (nativity and locality), and the other associates birth with poetry (nativity and genius); the one attaches Clare to his body, to a particular time and place; the other—in no uncertain terms—conceives of poetry not as a form of vocation, but a form of life and transcendence. Romanticism, especially in the familiar forms articulated by Wordsworth and Coleridge, can be understood as conjoining these two notions of birth, even as they are exactly the conjunction that Clare sought (and failed) to overcome when he sketched his own grave. The grave, in other words, makes poetry and posterity a matter of biology, that is, a matter of life itself.

Poetry understood in this way, far from a romantic innovation, is instead part of a much longer tradition that at this moment converges with a newly biologized notion of the political subject. The first known use of the phrase "a poet is born not made" dates to Pseudo-Arco, in a commentary on Horace's *Ars Poetica* from around the seventh century. In the *Ars Poetica*, Horace identifies in Democritus two ideas of the poet. The first holds that a poet's ability is "an inborn, not acquired talent or natural skill, and so depends on an internal condition that is constant." The other adds that poetry "comes from inspiration or a divine madness, and so depends on a force operating from without that is spasmodic." Now, it turns out that Democritus actually held only the latter position, that poetry is inspiration, frenzy, or *enthusiasmos*, not the former sense of genius and *phusis*, which arrived later, most likely with Pindar in the fifth century. So, the emergence of this phrase and its understanding of poetry as a natural talent, coincides with an archaic mistake, one that congealed in the late eighteenth century, and that has particular resonance for Clare (for whom frenzy and genius seemed to go together). While there is some evidence that Aristotle and Cicero may have combined the intrinsic and extrinsic (or natural and supernatural) theories of poetry, Horace's misreading, via his commentator Pseudo-Arco, can be traced as the source of this notion of the poet and poetry, which is found in Cervantes in 1615, and translated into English in 1620 ("For it is a true opinion, that a Poet is borne so, the meaning is, a poet is naturally borne a Poet from his mothers wombe" but then concludes, citing Ovid, "there is a God in us") and later in Edward Young (1756) who conceives of genius as both inspired and innate.[9]

The phrase "A poet is born not made" enters into English somewhat earlier than Spanish, in the 1570s and can be found in Lodge's *Defence*, Sidney's *Apologie*, and Webbe's *Discourse*, and then virtually disappears until Young evokes it, and Coleridge, reflecting upon Shakespeare's "genuine verse" in the *Biographia Literaria* writes: "But the sense of musical delight, with the power of producing it, is a gift of imagination; and this together with the power of reducing multitude into unity of effect, and of modifying a series of thoughts by some one predominant thought or feeling, may be cultivated and improved, but can never be learnt. It is in this sense that 'Poeta nascitur non fit.'"[10] So, for Coleridge there are two powers, one of perception and production, and one of unification; one of poesis and the other of synthesis. The Romantic conflation of the formerly opposed supernatural and natural conceptions of poetry figures natural genius as opposed not to the supernatural, as an extrinsic power of inspiration or enthusiasm, but to the nonnatural (mechanical, rhetorical, learned, contrived).[11] This is a difference between poetry and oratory (or rhetoric). Yet, just as Clare is embraced as a poet in whom natural and supernatural powers are fused, so too does the opposition between the natural and the contrived emerge as a new scene of the contamination of poetry and rhetoric, which is evident, for example, in the review of the *Village Minstral*.

When Clare's friends undertake to remember him, they presumably arrive at the phrase *Poeta nascitur non fit* not directly through Coleridge or Sidney, let alone its ancient sources, but indirectly, through the *Eclectic Review* or most likely through the first biography of Clare, Frederick Martin's 1865 *The Life of John Clare*. There, Martin writes:

> Of *Poeta nascitur non fit* there never was a truer instance than in the case of John Clare. Impossible to imagine circumstances and scenes apparently more adverse to poetic inspiration than those amidst which John Clare was placed at his birth. His parents were the poorest of the poor; their whole aim of life being engrossed by the one all-absorbing desire to gain food for their daily sustenance. They lived in a narrow wretched hut, low and dark, more like a prison than a human dwelling; and the hut stood in a dark, gloomy plain, covered with stagnant pools of water, and overhung by mists during the greater part of the year. Yet from out of these surroundings sprang a being to whom all life was golden, and all nature a breath of paradise. John Clare was a poet almost as soon as he awoke to consciousness.[12]

Here we see the conflation of the natural and supernatural (birth and inspiration). In Martin's account, the poverty of Clare's surroundings is the mark and evidence of his genius; however, what is unusual here, and crucial, is that rather than overcome these surroundings, he remains attached to them as their issue. Clare is both an outsider and a native son. Likewise, Cyrus Redding writes in 1841, having just visited Clare at the asylum from which he would soon escape: "I never before saw so characterized personally the *Poeta Nascitur* . . . which lifts men of genius above the herd."[13]

Read in the context of the very anxiety about fame and recognition that led Clare to feel despondent and close to death throughout much of his life, this grave is a mixed blessing. Indeed, it reflects his fame, success, and the appreciation of his followers who recognize him as both a local hero and an icon of genius; at the same time, it reflects a scene of misrecognition, where fame is reduced to the temporality of bodily life rather than that of untimely or transcendental poetry. It naturalizes Clare, rendering him a native genius, a poet by necessity rather than labor or will, one for whom bodily or biological life (birth) is indissociable from poetic or posthumous living on; and yet, the very gesture through which Clare is naturalized and through which birth is posited as the source of his creativity marks a break from the natural scene (rural life and poverty) through which he is made recognizable.

Clare's two graves—the one sketched in his journal and the actual monument—reflect two notions of poetry as it is related to biological or bodily life. The grave Clare describes in his journal anticipates the divisibility of the body from the work. It is a play on the two notions of corpus, where the burial of the physical body allows the literary body to live on, while the grave erected at St Botolphs, where Clare's body actually lies, announces their indivisibility. Clare's actual fame has relied upon the recognition of his poetry as an effect of bodily or biological life, even if it is this condition that Clare proposes to deny, perhaps not accidentally leading to a grave that marks the end of his hopes together with his ashes.

The epitaph's affirmation of the genuine or innate poet ("a poet is born not made") returns us to those accounts of political and personal sovereignty that Michel Foucault outlines in his 1976 lectures "Society Must Be Defended." In his discussion of the change in political sovereignty that occurred during the nineteenth century, Foucault claims that there has been a shift from one dissymmetry to another in the execution of power. If the right of the sovereign is a right over life and death, Foucault begins by noting that the relation to this right is not merely equivalent, as if life and death are treated in the same manner, whether figured as "giving" and "taking" or allowing to live and allowing to die. Rather, there is an initial incongruence in the very way that sovereign power is manifest, such that power obtains from *taking* life or *letting* live, the power to kill or not to kill. Yet, according to Foucault, in the nineteenth century this right is inverted, so that what determines political power is "the right to make live and to let die," and politics becomes biopolitics, a politics of the human race. In figuring "biopolitics," Foucault also points to a shift in preoccupation from "temporary disasters" (epidemics) to permanently living with death (endemics), from war to racism to poverty to climate change.

Clare's grave can be understood to emerge as a both a minor instance in this history of sovereignty and a disruption of it. Clare's experience of poverty and chronic illness, both mental and physical, led him to imagine a grave many years before his actual death. The inscription of this death, not on a stone, but in the journal also reflects back on the writing of life. Lauren Berlant has described "slow death" or diurnal death as a form of attrition in which living becomes a mode of dying. This notion of life become an unremarkable death would seem to differ from the possibility of living beyond death in posterity that renders biological death an unremarkable event in the face of literary survival. Clare's journal entry archives the convergence of two apparently opposed experiences of survival. Yet, by 1867 Clare's actual grave bears almost no trace of the radicality of this condition. Instead, it becomes an instance in the history of a biological or hereditary notion of poetic genius. It becomes a conventional, even citational, nineteenth-century operation. If Clare's 1824 journal entry invokes the multiple meanings and possibilities of survival in the division of biological life and poetic afterlife, his grave of 1867 exemplifies the fusion of poetry and biology.

Having said this, which is to say, having suggested that the poet typically associated with a certain natural history, also reflects a resistance to the biopolitical understanding that his work nevertheless seems to spawn and attract, I want to return to the journal entry in which Clare notes the importance of his "temporary conserns," before relegating them to the appendix. As I noted earlier, Clare begins the entry with an account of his health: "Very ill today and very unhappy my three Children are all unwell had a dismal dream of being in hell." Yet, that entry concludes with a rather different account of life and death, nature and survival, one that is allegorical and seems an antidote to Clare's commemoration as "born not made" and the Northamptonshire peasant poet. Clare writes: "Neglect is the rust of life that eateth it away and layeth the best of minds fallow and maketh them desert—Done nothing."[14]

While Frederick Martin, in the passage that leads to the inscription on Clare's actual grave, calls Clare "a being to whom all life was golden, and all nature a breath of paradise," in the journal entry that concludes with Clare describing his imagined grave, *life*, far from golden, is a base metal, subject to corrosion not through overexposure to the elements, but neglect by the public. Abandonment, disinterest, forgetting of the sort that Clare describes in "I Am" ("I am, yet what I am none cares or knows;/ my friends forsake me like a memory lost"), and the experience of spending his middle and late years in an asylum all reflect a form of wasting away, a life that for all of its poetic productivity is a slow death to which the journal becomes a daily commemoration.

If forgetting is figured as an act of destruction, can remembrance too be a form of neglect? Can valorization, the attention that in some respect Clare longed for when he figured himself a boxer or a Byron, be a version of disregard? If this is the case—and if this is what one reading of Clare's actual grave reveals in its difference from his early wishes, it is not the end of the story. The transformation of poetic life into biological life assumed by the grave at St Botolphs and the history of poetry in which it participates is threatened by the possibility that this remembrance also might be a form of neglect and that it might corrode the very life that it posits. This is what we discover when we read Clare's physical grave according to his account of destruction. When fame and memorialization conjoin biological and poetic life, they prove an antithetical mode of neglect, and memorialization—not only the work in the fields and illness and the later experience of the asylum—becomes a slow death that "eats away" at life and that leaves the mind fallow. One way of describing this grave might be to borrow Clare's words, the words with which he concludes this entry (and also the previous one): "Done nothing." In the stripped-down language of inoperativity that we expect of Samuel Beckett rather than John Keats, W. H. Auden rather than Jean-Paul Sartre, we find a description not only of Clare's ongoing life, but also of his grave.

At this moment, the quotidian and the monumental, the divided and the conflated seem to arrive at the same end. How could they not? Clare aspired to a future that depended upon the negation *and* affirmation of his biological life in the service of poetic survival; and his public construed of poetry as a biological inheritance, albeit one that reflects not continuity with nature, but its transcendence.

Although Clare's plans for his grave may be prompted by a rather serious bout with the flu, the sketch of the grave that appears in his journal is more than just an occasional musing, but connects closely to the premise with which Clare began. On the opening page of the journal, just opposite the title page, Clare also sketched a grave. This stone bears not his name, but the "naked" names (to recall Wordsworth on epitaphs) of three other poets: Chatterton, Bloomfield, and Keats. Jonathan Bate understands this gesture as an act through which Clare "place[s] himself firmly in the prematurely Dead Lower-Class Poets' Society," indeed Bloomfield's recent death upset Clare deeply.[15] Yet, this collective grave of natural genius and poetic posterity is more complex than Bate's description implies. These are dead poets who live on through their poetry. Their grave, at least in Clare's sketch, is undated, and their survival contingent on the effacement of the very biological life that heightened their fame. Clare does not simply list his name alongside theirs, just as he did not simply leave his grave undated. Rather, Clare writes his *life*—his everyday life, not just his name—alongside their deaths. September 6, the day upon which the journal begins, is one that Clare calls "a beginning," and which we can understand as the beginning of a quotidian textualization of life that introduces the possibility and promise of life beyond life (posterity), not in the prestige of poetry, but in the everydayness of the journal. At the same time, this life beyond life bears, as its condition and despite Clare's wishes, biological life: the life of the peasant body whose ailments and sufferings cannot—will not—be arrested by the image of a grave and whose posthumous existence continues to be guaranteed not simply by the existence of so many pages of poetry, but by the recollection of the peasant body in its suffering, corrosion, and discontent, that is, a life as unremarkable death. It is this paradox, this unresolved experience of the relation of life to poetry that describes not (or not only) the operations of a biopolitical epoch, but the unaccounted work of romantic biopoetics, the gesture through which this life comes to be borne.

3. Can the Poet Speak?

When John Ashbery received the Charles Eliot Norton Professorship of Poetry at Harvard in 1989 and was asked to deliver six public lectures, he used the occasion to reflect upon the "other traditions" that have informed his work. By "other traditions," which is also the title of the published version of the lectures, Ashbery refers not only to an alternative canon but also, as he puts it, to one made up of the poets that he "reads habitually in order to get started; a poetic jump-start for times when the batteries have run down."[1] First among these energizing, animating poets is John Clare. Clare's poetry works on Ashbery as a kind of calling that sets him back on track: its effect is "always the same—that of reinserting me in my present, of re-establishing the 'now.'"[2] Understood in this way, Clare's poetry operates like a rhetorical figure—a prosopopoeia, or fictional apostrophe—that returns the possibility of poetic production and of the writing self.

As if registering Clare's effect on him, Ashbery opens the lecture on "John Clare" within that space of self-presence. He wonders, with some modesty, what his audience might expect of the Norton lectures and speculates that he was invited to give them in order to reveal the meaning of his "hermetic" poetry, something that he insists he will not—and cannot—do. As he explains it, "There seems to be a feeling in the academic world that there's something interesting about my poetry, though there's little agreement as to its ultimate worth and considerable confusion about what—if anything—it means," and he goes on to disabuse his audience of any hope that this clarification might be forthcoming. This is because, as he explains his thought "is both poetry and the attempt to explain that poetry; the two cannot be disentangled."[3]

When Ashbery claims that his poetry and thought are indissociable, he seems to exemplify a lost notion of poetry and thinking that Giorgio Agamben describes in a short essay on "Creation and Salvation." There, Agamben reflects upon the shared origins of philosophy and poetry, laments their postromantic division, and suggests, somewhat nostalgically, that we live in a fallen world in which poets and critics have become divided from one another. It is a world in which poetry needs criticism for its meaning, and criticism needs poetry for its object. On the one hand, Ashbery, in offering nothing other than poetry as the meaning of his work concedes that it is not the poet, but the critic who "saves" the work from its hermetic existence and invents its meaning. He (the poet), having "invented" his poetry, has nothing more to say. On the other hand, Ashbery claims that his poetry already is thought. All of his thought is there in the poem, so that the very intimacy with thought borne by his conception of poetry renders his work already salvaged, even if that salvation remains a cryptic one.

While Ashbery's refusal to explicate his work, leaves him reflecting on its conditions of possibility, for Clare the division between *creation* and *salvation* is fraught from the outset, insofar as his agency as a poet is permanently bound up with the editorial process and with salvation in another sense. Clare's editors and readers, like Tom Paulin, who writes that "it could be that Clare—shy, feral, intensely gifted—will never be redeemed from all the neglect and mutilation he has suffered" espouse a salvation that only they—if anyone—can achieve. Paulin's account of Clare, which for the moment will stand in for all of the other competing accounts that nevertheless share its logic raises fundamental questions about poetry as a form of life and editing as an act of salvation—or destruction.

Since John Taylor's first edition of Clare's *Poems Descriptive of Rural Life and Scenery* (1820), the question of how to present Clare's poetry has obsessed scholars and critics. Over the years, they have struggled with questions like: Should Clare be considered a romantic poet, or does this categorization lead us only to "colonize" Clare rather than to recognize him as the misfit that he appeared to be during his infrequent trips to London? Should we read Clare's asylum poetry—including poems that he wrote in Lord Byron's name—as continuous with his early poetry or rather as the "poetry of madness," whether an aberration or a lens through which Clare's complete works can retrospectively be understood?[4] Yet, it is the question of *editorial* presentation, specifically, whether Clare's language, spelling, and punctuation should be "normalized" or published as they appeared in manuscript (or as close to that appearance as possible), that has attracted the most intense and ongoing attention.

Clare's first editor, John Taylor—who was also Keats's publisher—was a great supporter of Clare, even though he was well aware of the difficulty of finding a significant audience for the poetry. Taylor was responsible for publishing the *Poems Descriptive of Rural Life and Scenery*, yet there is little agreement on the merits of Taylor's editorial role. Tim Chilcott suggests that from at least 1821, Clare "began to rely increasingly on Taylor to transform rambling and untidy manuscripts into poetry ready for publication."[5] The result, in the words of another editor, Arthur Symons, is that "it is difficult to know how much of the early poems were tinkered for publication by the too fastidious publisher Mr. Taylor."[6]

Eric Robinson and Geoffrey Summerfield follow this line of inquiry and aim to expose "the nature and extent of Taylor's 'slashing' of *The Shepherd's Calendar*" (the word "slashing" is, as they acknowledge, Taylor's own).[7] They go on to argue that even "when all appropriate allowances have been made, Taylor cannot be judged a consistently reliable editor."[8] Jonathan Bate points out that Taylor did precisely too much "tidying" and leaves Clare's poems not cleaned up, but positively "botched."[9] It is this tidying up that led Robinson and his coeditors to undertake the massive project of producing a nine-volume collected Clare for Oxford, a collection of poetry and prose that aims to present Clare's manuscripts "intact."

Robinson goes so far as to call Taylor "careless, dilatory, bullying" and insists that "it must be appreciated that Taylor was not simply trying to transcribe Clare's manuscripts, a difficult task in itself, but also to alter Clare's vocabulary, grammar, spelling, punctuation, and sometimes sentiments."[10] In undertaking to correct Taylor's excesses, which Robinson figures as intentionally malicious, the Oxford volume intends a restoration and transcription (rather than what we might call a translation) of Clare's original manuscripts.[11] As a result, the volume maintains the obscure and sometimes simply mistaken spelling and grammar found in Clare's manuscripts, reproduces Clare's alternatively minimal, excessive, or incorrect punctuation, and tries to recover poems that have been erased, written over, and muddled.[12] Bate, while sympathetic, to the aspirations of this project also calls this process not editing, but "unediting," and he compares it with "the restoration of an over-varnished canvas to its original colours."[13] He acknowledges that the effect of such restoration is that "Clare's language was revealed in all its freshness and immediacy," but also worries that the confusion spawned by poor punctuation and misspellings also can interfere with the ease—and even immediacy—of the poems.[14]

In the introduction to his 1873 volume of Clare's life and works, J. L. Cherry explains that "of those which are printed, scarcely one was found in a state in which it could be submitted to the public without more or less of revision and correction."[15] Cherry holds the position that Clare must be edited, that his manuscripts are incomplete or filled with errors that it is the duty of an editor to correct. Bate's position resonates with this one, although he is clear to distinguish between acts of correction and the kinds of rewriting that Taylor undertook. Alternatively, in his 1908 edition of the poems, Arthur Symons concedes: "I have tried in vain to find the original manuscripts, which I would have liked to have printed exactly as they were written, having convinced myself that for the most part what Clare wrote actually was better than what his editors made him write."[16] Robinson and his coeditors share in this conviction and go to extraordinary lengths to reproduce the manuscript as closely as possible believing that any alteration is just that, an alteration. Yet, even Robinson, who has claimed copyright of the poems, admits that while he can preserve Clare's text, he cannot reproduce the experience of encountering the manuscript, which would require transmitting the smell of vinegar, like a chip shop, that lingers on them.[17]

Despite ideological and stylistic differences, these editors, past and present, share in the wish to let Clare speak and be heard. In this sense, they differ not in their aims or in their assumptions about the task of editing and publication, but only in their means. Although Bate compares Robinson's editorial project to the restoration of a canvas to its original state and thus sees the Oxford texts not as worn out, but too enhanced, Bate is like Robinson, and for that matter, like Taylor himself, in that he is involved in his own project of restoration, not of an image (the well-worn image of Clare as peasant poet), but rather of voice, a voice that could be audible, authoritative, and true. Thus, it is the question of *how* this amplification can be accomplished, rather than *whether* and *at what cost* such a project is possible, that marks their differences. Indeed, this question of voice leads Simon Kövesi to cast the two editorial camps as figures in a political struggle and to argue that debates about how to edit Clare mirror debates about standard language in Britain (which he figures chromatically as "red" and "blue," colors that have opposite meanings in Britain and the United States), explaining thus why this scholarly difference has attracted the attention of leftist thinkers like Paulin.[18] Yet, these positions need not be read only as symptomatic of contemporary politics. Rather, they can be analyzed as evidence of the confusion of aesthetic and ethical positions in literary scholarship and editing. Whether one can simply overcome this confusion or whether, as the most visible examples seem to suggest, the aesthetic and ethical projects—the supposed restoration of Clare's voice and the presentation of Clare's text—are inextricably linked to the project of publication itself is the question that I take up in the pages that follow.

Put another way, the question of editorial presentation has both ethical and aesthetic dimensions, yet the debates about how to edit Clare reveal that an aesthetic or figural act is the effaced ground of the editors' ethical claims. Arguments about how to edit Clare assume that it is more or less ethical to allow Clare's "true" voice to be heard for the first time by publishing it without editorial interference than it is to rescue Clare from the illegibility of his most faithful editors (those who maintain his poor spelling and absent punctuation) and thus render him accessible. No position within these debates can do without positing Clare's authentic voice as the ground for their claim. Their ethical force depends upon this fiction of voice, this act of presentation. Whatever their differences, insofar as the debates surrounding Clare's manuscripts rely upon ethical positions, that are also aesthetic positions (or presentations of Clare), they are examples of aesthetic ideology at work. They reveal an ethics of reading (and editing) that rests upon an aestheticization of their object (Clare), which is also the very possibility that their truth claims must deny.[19]

In 2003, Jonathan Bate published not only a new (and intentionally popularizing) edition of Clare's poetry, but also a comprehensive biography of the poet. Taken together, these two volumes revived interest in Clare and his work; yet, the coincidence of these two publications is also important. Many critics, including Sarah Zimmerman, have noted that Clare's biography has regularly been read in lieu of his poetry, and the author's life, "a young peasant, a day-labourer in husbandry, who has no advantages of education beyond others of his class," as Taylor explained in his first edition of Clare's works, has been of utmost importance in the reception of Clare.[20] In some respects, Bate's double-barreled publication project can be understood as a *response* to this tendency. By issuing a collection of poems alongside the biography, by coordinating their covers and including cross-references, Bate registers the necessity of reading the life and the work together. At the same time, the copublication of poetry and biography also seems to repeat the structure that Zimmerman identifies, for Clare's biography once again overwhelms his poetry in size and scale.[21] The biography is so hefty in part because it promises to be authoritative and free from error in distinction from Clare's earlier biographers, above all the first, Frederick Martin, on whom Cherry also bases the *Life and Remains*. Just as Bate envisions himself as being true to Clare's life, he also describes his editorial works as following Clare's intentions by adhering to a "middle way," one that "avoid[s] errors and alteration, but provide[s] light punctuation and regularize[s] the spelling without diluting the dialect voice."[22] Bate figures the question of editing as an ethical question. He asks, "Is justice done to him by the presentation of his manuscripts in the raw, free from the shackles of prescriptive grammar?

Or does the reproduction of their idiosyncrasies unintentionally perpetuate the image of him as a semi-literate primitive, an eternal child?"[23] In order to answer this somewhat hyperbolic question, Bate turns to Clare and explains: "Clare indicated in a note to his publishers that he expected his editors to normalize his spelling ("I'm" for "Im," "used" for "usd," etc.) and to introduce punctuation for the sake of clarity, but he did not want them to over-regularise his grammar or remove the regional dialect words that were so essential to his voice."[24] Bate does not cite Clare's note in the introduction to the selected poetry, but it does appear in the biography. Indeed, upon having described the Oxford edition as "stripped of all punctuation and replete with misspellings, slips of the pen and so forth," he asks "Is this what Clare would have wanted?"[25]

Bate then goes on to cite parts of the note Clare wrote to his publisher in which Clare "explained that he had not attempted his own '*Stops or Punctuation*' and that 'Bad spelling may be corrected by the amanuensis, but no word is to be altered.'" Bate interprets the passage to mean, "His clear implication is that stops and spellings *should* be corrected by the professional scribe, but that his lexical choices are to be respected."[26] Thus, Bate justifies his process as a faithful one, while, at the same time, he betrays it by not lowering the "s" or the "p" in "*Stops and Punctuation*." This point, however, is incidental to the larger suggestion that I am at work to make, which is that Bate promises to do justice to Clare by actually listening to him, by finding a text addressed to another and understanding it as addressed to him. In this respect, the authority of the editorial project, like that of the biographical project, necessarily rests upon speculation and ventriloquism. Thus, despite his active intervention in debates about *how* the poetry should be edited, Bate cannot settle finally the question of the proper nature of that poetry, and whether it should be served, to use his own metaphor "raw or cooked." The most obvious reason for this ambiguity is that his claim relies upon the same evidence as Robinson's claim.[27]

Several critics already have attested to the limitations of or contradictions within Bate's project. For example, in his review of Bate's *Selected Poems of John Clare*, a British edition of the collection, R. K. R. Thornton notes that, far from uniformly consistent, Bate's editorial project relies upon a projective identification of Clare's "spirit" and promises to correct "alterations [by previous editors] that go against [it]."[28] Thornton worries that "alternations of [Clare's] text can be corrected, but can his spirit be agreed upon?"[29] Yet, by framing his question in this way, Thornton also seems to suggest that, unlike a "spirit," Clare's text could be "corrected" and hence agreed upon. However, this is precisely *not* the case, for Clare's text as much as his spirit, perhaps even more than his spirit, remains in question. Likewise, in an essay that imagines a "green edition" of Clare, a project that no doubt is indebted to Bate as one of the vanguard of literary ecocriticism, Kövesi argues that "Bate's return to the manuscripts to edit them with his "'new-found' fundamental principle [i.e., following Clare's wishes to the letter] is also a species of 'textual primitivism.'"[30] Moreover, Kövesi goes on to suggest that even Bate cannot always abide by his own rules and includes some "raw" texts when "the problems of regularizing seem too great."[31]

One way of describing this situation is to recognize that the ethical aim of these editorial projects ("doing justice to Clare") is constituted by a single, recurring gesture, that is, by a rhetorical figure, a prosopopoeia, the master trope of the lyric (as well as of autobiography and witnessing), through which the mute, dead, or absent are made to live and speak.[32] Prosopopoeia in this instance does not solve the problem of absence or death, but ultimately calls attention to its impasse. In this case, Clare's muteness, his failure to communicate, reemerges at the very moment that his editors insist upon having heard him properly and having allowed him at last to speak. When the editorial horizon becomes the restoration of Clare's voice, the ethical project becomes indissociable from a figural one.[33]

Now this insight—the entwinement of the ethical and the figural—is far from original. In fact, J. Hillis Miller has shown that "doing and evaluating are related to personification," and "doing and evaluating" is precisely Miller's way of signifying "ethics."[34] But what is distinctive in Clare's case is the necessity of a denial of figuration (or, stated positively, the necessary claim of authenticity) at the moment that an ethical determination rests upon a prosopopoeia for its force. In other words, Clare's editors gain their legitimacy by a claim to authority that retrospectively produces the voice and the subject that they figure as speaking to—but also through—them. In this sense, the editorial projects rely upon a fiction of posthumous address. This fictional voicing at once marks the difference between these projects (Bate and Robinson hear Clare differently) and collapses this difference by virtue of depending upon a fiction for their authenticity.

In the appendix on "Clare's Text" with which he concludes the biography, Bate marshals a version of prosopopoeia in which the dead (Clare) is understood as living and speaking. Bate writes (and these are the book's last sentences): "The John Clare Society thrives. Schoolteachers have discovered that his 'nature' writing is an ideal way to introduce children to poetry. Enthusiasm for the work and fascination with the life may be found among a startling array of common readers, many of them far beyond the academic environment in which poetry is often confined. At long last, John

Clare is in good health. But we still await a balanced presentation of the full range of his texts."[35] Here, Clare becomes "healthy" for the first time, a metaphorical health and biologization of his poetic life that evokes several senses of illness: Clare's physical and mental illness in his life and his neglect in life and after death. This health is signified by an interested nonacademic readership, a thriving society devoted to his work, and a place in the school curriculum, and it occurs despite ongoing uncertainty about the very texts through which Clare continues to "live."

The appendix suggests that this work draws its legitimacy from listening to Clare, but it is also the condition of the very voice upon which this legitimacy rests. This metaleptical structure, a confusion or substitution of causes and effects, far from resolving the problem of Clare's life and death, health or illness, instead repeats it. Moreover, if Bate seems to suggest that Clare will live despite his texts' mutilation, this suggestion calls into question the importance of the debates in which Bate has been a central participant. Or, understood another way, it is precisely *because* Clare is still speaking (because he is in "good health") that we must listen to him, but he speaks and is healthy only through these texts.

When Bate describes the Oxford editorial project, he does so by explaining that Robinson "believed he was restoring the *authentic* voice of Clare, freeing him from the kind of control that had been exercised over his work by Taylor and the other early editors."[36] Rather than turn away from a logic of restoration and authenticity, Bate offers only another version of it when he undertakes to present Clare in such a way that he might be read and received. In other words, Bate's complaint about Robinson is not that he assumes that the authentic voice can be restored, but rather that he understands the authentic voice to be registered through poems transcribed directly from the manuscript, blemishes, errors, omissions and all. Bate turns to Clare in order to justify this concern and recalls that "Clare knew that he wrote too profusely and that he needed editing" and Bate insists further that the "key" to "Clare's art" is "immediacy," which, in his view is ironically unobtainable when one must first work through a manuscript filled with unfamiliar spellings, awkward punctuation, and other distractions.[37] Thus, Bate's project, like Robinson's is underwritten by an attempt to reveal Clare's true art by listening to him, yet Bate undertakes to achieve this by shaping and "drilling" (Clare's term for punctuating) without tidying and botching.[38] This process appears to restore that which was lost, but at the same time, in figuring its project as a restoration it also denies the absence of authoritative poems.[39] This is not to say that there are no poems by Clare, but rather that we cannot finally establish how a poem by Clare should appear.

In this sense, it is not merely a question of agreeing on Clare's spirit, but rather agreeing on the order, shape, and existence of his texts. The problem of appearance is so fundamental that we cannot conceptualize the project as one of restoration without betraying the crisis of appearances that is the context for these debates. Indeed, the framework in which editors have staked out their positions as ethical positions (or questions of justice) cannot but rely upon the trope of the restoration of a voice. Thus, they edit the poetry in a way that assumes a listening to Clare while at the same time establishing the vehicle through which he might "speak to us" and "be heard." These gestures, in making claims to authenticity, also invent a textual past that never quite existed, and yet they pass this invention off as an act of faithfulness and justice.

In an essay on the limits of trauma theory, Dominick LaCapra reflects upon the distinction between *absence* and *loss*, a difference that is also could be understood as central to Paul de Man's account of prosopopoeia.[40] While LaCapra recognizes that the difference between *absence* and *loss* is important for reasons of "intellectual clarity and cogency," he is ultimately concerned with the "ethical and political dimensions" of this distinction.[41] In his account, loss is an historical category whereas absence is a transhistorical one. To confuse or conflate the two, to assume that what once existed has never existed or to assume that one could recover what has never been the case as if it was merely lost, can lead to myth, nostalgia, or melancholia.

In LaCapra's terms, Robinson and Bate both can be understood to construe Clare's poetry (and voice) as having been *lost* in the work of prior editions and to frame their own editorial projects—whatever their significant differences—as the restoration of a text, and implicitly a voice, assumed to be lost. The editors both abide by a belief that their task is to recover this voice in such a way that it might properly be heard, and perhaps even be heard for the very first time. Yet, the question remains: Has something been lost—or was it never there in the first place? And if it never existed or appeared, or never existed as a unity, if there ultimately is no proper, authentic manuscript or way of receiving the manuscript, only a multiplicity of pages and directives and practices that cannot finally be resolved into one authoritative text or voice, do these editorial models encounter the same limit: the confusion of absence and loss in the claim to restore or recover what never was lost because it never was there to be lost. This confusion, which also corresponds to the grounding of ethics in aesthetics, or the positing of voice as the condition of authenticity, can be seen to complicate LaCapra's account while exemplifying it. For, to speak of Clare's authentic voice as absent may not be the same as saying that there is no poetry. This absence is not the same as the romantic absence of work that Maurice Blanchot describes when he acknowledges that for all their talk of poetry, the Jena romantics actually failed to write poems, for Clare wrote plenty of poems—hundreds and hundreds of them. Yet, this excess of poems and the open question of poetic completion correlates to a form of absence that easily can be misrecognized as merely a distortion or disappearance: the nonidentity of Clare's work that is spawned by indeterminacy, incoherence, and uncertainty.

LaCapra recognizes that the conflation of absence and loss has both affective and political implications. As he explains: "When absence is converted into loss, one increases the likelihood of misplaced nostalgia or utopian politics in quest of a new totality or fully unified community."[42] LaCapra goes on to suggest that "the difference (or nonidentity) between absence and loss is often elided, and the two are conflated with confusing and dubious results. This conflation tends to take place so rapidly that it escapes notice and seems natural or necessary," and he explains further that "the conversion of absence into loss gives rise to both Christian and oedipal stories (the Fall and the primal crime)—stories that are very similar in structure and import."[43] It may seem that we have traveled quite far from Clare and his editors. And yet, it is precisely the nostalgic, naïve, antihistorical aestheticization of the political that LaCapra recognizes as the outcome of this conflation. While recent biographies and introductions have revealed Clare's awareness of his poetic predecessors (as well as his awareness of local ballads, etc.), which is to say the extent to which he was not quite as naïve as his popular image as "the peasant poet" suggests, it might be that the process of editing, even when, as in Bate's case, it aspires towards nuance and avoids valorization of the primitive, nevertheless reintroduces the very logic it claims to reject.

This confusion of absence and loss has still a further manifestation here. Clare's editors, each in their own way, focuses not the absence of voice or life, but rather its *loss* or destruction by their predecessors and peers. They construe the editorial project as an act of recovery, albeit one that allows Clare to be heard for the first time. Yet, conceiving of editorial work in this way also suggests that it is an invention (production of something that never before existed) rather than restoration (recovery of a lost or mutilated text), raising questions about the nature of authorship and authority, which they claim to defend. Indeed, the difference between recovery and invention is disturbed when restoration is understood to allow for a belated, if nevertheless originary experience. One result of this concept of restoration as allowing for an originary event is that Robinson's claim to hold the copyright on Clare's texts, rather than appear as violent and inappropriate, can be seen to reflect the impact and the stakes of editing. Understood in this way, editing *is* a mode of production or invention, rather that mere coordination.

A reading of Clare's "I Am," a poem that, despite its ambiguities, raises the question of voice and life in a remarkably clear fashion, might give some insight into these editorial—and textual—predicaments. Indeed, the poem reflects on loss, self-loss above all, and forgetting, while also evoking the utopic and nostalgic tropes that LaCapra describes. As a poem that opens in a lament for absent recognition, "I Am" might tell us something about the debates surrounding the presentation of Clare's texts, just as it reflects (and effects) one of the central ambiguities in Clare's reception: the confusion of textual and personal identity. Moreover, as a poem that appears in both Robinson's and Bate's editions (as well as having been published in at least seven other editions), it is also an episode within the debate. Here is the text of the poem as it appears in the Oxford edition:

1

I am—yet what I am, none cares or knows;
My friends forsake me like a memory lost:—
I am the self-consumer of my woes;—
They rise and vanish in oblivion's host,
Like shadows in love's frenzied stifled throes:—
And yet I am, and live—like vapours tost

2

Into the nothingness of scorn and noise,—
Into the living sea of waking dreams,
Where there is neither sense of life or joys,
But the vast shipwreck of my lifes esteems;
Even the dearest, that I love the best
Are strange—nay, rather stranger than the rest.

3

I long for scenes, where man hath never trod
A place where woman never smiled or wept
There to abide with my Creator, God;
And sleep as I in childhood, sweetly slept,
Untroubling, and untroubled where I lie,
The grass below—above the vaulted sky.[44]

Clare's "I Am" accounts for an experience of suffering that approaches melancholia. It opens with an interrupted statement, "I am—yet what I am none cares or knows," and uses a series of comparisons to describe the failed recognition of both the subject ("My friends forsake me like a memory lost") and his "woes" ("They rise and vanish in oblivion's host/Like shadows in love-frenzied stifled throes"), which both determine and evacuate him. Understood in this way, the poem relies upon figures of comparison to stage a scene of recognition that does not take place outside of it.

One could say that the ongoing debates surrounding the publication of Clare's texts point to a situation opposite of the one described in the poem. Clare's editors harbor an almost obsessive desire to know Clare, and they defensively care for him, protecting him from further disappearance or mutilation. Yet, their insistence upon the authentic Clare issues through an act of figuration. The prosopopoeia that I have discussed, rather than an instance of adequate recognition, can be seen instead as a form of effacement and misrecognition. What appears to be an act of restoration (making Clare speak) is instead an act of invention, albeit one that must deny its status as such. Thus, the obsession with Clare can be understood both as a response to and repetition of this nonknowledge, that is, as another instance of the impossibility of knowing *what* Clare is, rather than a remedy for the abandonment that the poem describes.

At the same time, as Timothy Morton has pointed out, the poem can be read autobiographically in another sense. For Morton "I Am," as if anticipating Mallarmé and Ashbery (or recalling medieval riddles) accounts for its *own* existence as a "spectral quasi-object suspended in nothingness, an inconsistent bunch of squiggles that cannot ever know itself as such."[45] Understood in this way, far from recovering a human subject in pain, the poem instead refers only to itself. It is an instance of specular self-reference and nonunderstanding that comes into being through a statement of obscurity, and as such allegorizes poetic emergence. Following upon Morton's reading of the tension between positing ("I am") and cognition ("none cares or knows"), between the lyrical-autobiographical subject (the *I* as authorial or poetic subject) and the poem as at once the subject of positing and the object of indifference and noncognition, we can see that the subjective predicament (the failure of recognition) emerges as indistinguishable from a poetic predicament. Indeed, this coupling of the subjective and the poetic, of the obscurity and betrayal of the subject and the obscurity and betrayal of the poem, is not far from what occurs in the ongoing debate about how to edit Clare. The very obscurity and indifference that the poem identifies, laments, and even repeats in the poetic act may be what the anthropomorphizing claims of editorial practices that rely upon "hearing" and "understanding" Clare also elide—and in eliding (in failing to recognize) repeat.

This repetition, a repetition of nonrecognition in the mode of restoration or recognition, is also reflected, if not explicitly thematized, in Clare's poem. While the poem seems to follow a path from ceaseless, even self-perpetuating, pain to an alternative world arrived at in the poetic imagination, that path ultimately has mere repetition, rather than hoped for freedom, as its end. For Clare, like his editors, this repetition also involves a confusion of absence and loss, invention and restoration.

Clare renders the drama of absence and loss in the poem's third and final stanza, which describes the first-person subject's desire for a utopic scene "where man hath never trod." Rather than reflecting a world that never could have been experienced before (a world in which this subject, like all human beings, is absent), this scene instead is figured as a return of what has been lost: childhood. What never could have been possible (the experience of a world without other men) and would need to be invented is understood as what already has taken place but no longer remains (childhood). The subject imagines and wishes for a place that has no human history, while at the same time figuring this place that by definition precludes him as the site of a return—the place of childhood and of oneness with his creator. The stanza reads:

I long for scenes, where man hath never trod
  A place where woman never smiled or wept
There to abide with my Creator, God;
  And sleep as I in childhood, sweetly slept,
Untroubling, and untroubled where I lie,
The grass below—above the vaulted sky.

This longed-for "scene" of unity with God in the absence of (or prior to) human presence also would be a return that would allow the subject to again experience the tranquility of childhood sleep.

The "scenes" without others, and hence without the abandonment and betrayal by them that the poem's first two stanzas describe, would be a paradise. They would belong to a world in which loss and woe would not be possible, while at the same time admitting that such a world would have to do without human life. Here, the conflation of an impossible "scene" (the "scene" in which he is the first and only human being) and a lost "scene" (the "scene" of childhood sleep), also suggests that childhood never will have occurred, like a world without humans that nevertheless could be witnessed and experienced by a first person subject, whether understood as man or poem. Childhood thus becomes an impossible or mythic prehistory, rather than a stage of life.[46] The poem concludes with the subject asleep and free from the pain described in the beginning of the poem. Yet, if the first stanzas describe a world in which "what I am, none cares or knows," the final stanza, with its turn away from present pain into a mythic past, turns out to be a repetition, rather than a radical change. If in the first stanza the poet is painfully ignored and invisible, in the final stanza he becomes not an object of care, but of irrelevance: "Untroubling, and untroubled where I lie,/The grass below—above, the vaulted sky." This mythic place, a new world figured as the recovery of a lost one, turns out finally to be merely the repetition of what is: oblivion, indifference, and abandonment.[47]

While within the context of the poem's itinerary, this blissful scene may be only another repetition of the same, evidence of the failure to break out of the world in mind or poetry, rather than a final arrival in paradise, it also stands as a figure of poetic vision in another sense. When Clare refers to the sky as a "vault" ("The grass below—above the vaulted sky"), he seems to see the sky as Kant, distinguishing between moral and aesthetic judgment, explains that poets do. Indeed, in a section of the analytic of the sublime whose antithetical logic Paul de Man has explicated in detail, what Kant perceives as a poetic way of seeing, is just that, a matter of seeing rather than thinking or interpretation or understanding. Kant's example is another vaulted sky: "If, then, we call the sight of the starry sky *sublime*, we must not place at the foundation of judgment concepts of worlds inhabited by rational beings and regard the bright points, with which we see the space above us filled, as their suns moving in circles purposively fixed with reference to them; but we must regard it, just as we see it, as a distant, all-embracing vault. Only under such a representation can we range that sublimity that a pure aesthetic judgment ascribes to this object . . . we must regard it as the poets do, merely by what the eye reveals."[48]

To see the sky as a vault is not to understand it as a distant homeland or a figural tomb, but rather as a mere form. This kind of seeing ("just as we see it"), which de Man associates with "flatness" and the absence of personifying gestures, does not involve understanding, imagination, or integration, but is an instance of Kant's formal—or as de Man translates it, material—sublime. While de Man turns to Wordsworth, as a poet who employs the same figure of the sky as vault, he shows that Wordsworth's use of the figure is precisely *not* poetic in the way that Kant understands it, and by registering this difference, he also points to a break between romantic and idealist, poetic and philosophical modes of seeing. For Wordsworth, the sky's sublimity rests upon its functioning as a shelter. It both protects and can expose the fragility of man, and this is the mark of its power. As de Man explains, for Wordsworth "the sky is originally conceived as a roof or vault that shelters us by anchoring us in the world, standing on a horizontal plane, *under* the sky, reassuringly stabilized by the weight of our own gravity."[49] De Man goes on, however to remind us that "if the sky suddenly separates from the earth and is no longer, in Wordsworth's terms, a sky of earth [Wordsworth writes "The sky was not a sky / Of earth"], we lose all feeling of stability and start to fall, so to speak, skyward, away from gravity."[50] And it is this sheltering, subjectivizing capacity that, in de Man's account registers the difference between two apparently similar claims: in Kant the sky only appears (and appears as a vault); it cannot be understood, however, to function as a shelter. Whereas in Wordsworth, we have not simply a matter of appearance, but an operation, yet one that does not merely domesticate the sky, but threatens the physical world.

In Wordsworth's autobiographical poem, the subject is at stake. Yet in Kant, despite the centrality of the subject to his account of the sublime, far from personifying, in this account of aesthetic judgment, we have a "flat" and formal scene in which there can be no mind, no consciousness, just pure "unmediated" vision. An aesthetic vision, as Kant imagines it, a vision of the sky as vault (or the ocean as mirror or abyss), is a purely formal, nonanthropomorphic, nonlyrical, or two-dimensional way of seeing. De Man shows in his succinct reading of Wordsworth that precisely what grounds the human also can unground or undo him, leading to a catastrophe of major proportions. Returning now to the moment in Clare's poem of self-positing, if not self-reflection, a poem noted for *its* sublimity, we are left with the question of how to understand the final line (and stanza) in which the sky emerges as a vault. Is this poetic vision like Kant's or like Wordsworth's? Is sublimity marked by flatness and form or by mind and imagination? Is this an aesthetic or a moral judgment? How does the poet see it?

In some respect, Clare's autobiographical poem resembles Wordsworth's autobiographical poem. And yet, for Clare the inversion of earth and sky is a matter of grammatical, rather than figural uncertainty. In the final line of "I Am," the subject, rather than standing upright, is instead already fallen: he is prone and asleep. While the dash between below and above ("The grass below—above the vaulted sky") seems to stabilize the relation, to hold apart and render parallel earth and sky, it is also like the "I" described in the poem, a horizontal rather than a vertical figure, one that is too weak to maintain the order of things. Rather, the blank separation—the dash—cannot secure the vertical relation between earth and sky and thus "the grass below" also can be heard and read catastrophically to stand "above the vaulted sky" (i.e., "the grass below [stands] above the vaulted sky").[51] While the subject may be "untroubling and untroubled," it would seem that such a scene is also nothing but trouble: trouble for the reader, trouble for the subject, trouble for the world.

If the poem at first glance might seem to end with the arrival at a utopic scene, a pleasant world of order and quiet, this world for all its thematic and textual resolution, nevertheless repeats the chaos and the disturbance of the poem's first two stanzas. In this sense, what appears to be the overcoming of pain through the imagination (i.e., in an act of mind) turns out rather to repeat and reveal the failure to recognize the self. Indeed, what had been cast as the fault of others here emerges as equally the doing of the self.

The world of which the subject dreams, neither absent nor lost, turns out to be merely a repetition of the same a world in which he endures, a world that even at the furthest reaches of the imagination he cannot escape. This inadvertent repetition can be understood traumatically as the compulsion to repeat or as an instance of the nondifference between life and death (this time cast in a negative mode, for death is one way of naming the desired state). It also suggests that whether imagined as an impossible scene or as the recovery of a lost scene, this utopia occasions an experience akin to the one that is the source of so much pain, living without recognition, without the capacity to be seen, heard, or understood. Thus, if there is a change in the poem, it is not due to recognition, but rather to numbness. The subject is not only "untroubling" but also "untroubled," unmoved, stable, blank, or flat like a dash, despite the catastrophe everywhere around him.

This stability, this being-untroubled, is also what the poem states with its insistence upon the "I Am" despite all else. The "I Am" is a statement of what remains untroubled, and in this sense it is a formal claim (Fichte, Coleridge), empty of content, and also a performative one.[52] But if "I am" becomes a statement of something like death, of total stability in the face of catastrophe (the world turned upside down) it is of course also a statement of the existence of the subject—however formal, empty, or iterable the statement may be. The vertical *I* and the horizontal dash, the upright self and the fallen, slumbering body *both* remain frozen, stable, and unmoved between heaven and earth.

Yet, for all of this, the lyric "I Am" ends in a grammatical crisis, rather than a figural one. Far from the sky as home or tomb, the sky (like the "I" with which it rhymes and that its rhyme repeats and incorporates) is a subject of grammar. The poem, which seems to rely upon personification for its address—the positing of self and other, reflection and mirroring in the absence of subjectivity and recognition—instead can be understood as an instance of the flatness or "depersonification" that Clare elsewhere describes. Thus, even a poem that would seem to involve poetic and autobiographical vision, a poem wrought with pathos—which is to say, even a lyric poem—here turns out to be a poem of nonpersonification, which is also one way of describing the very situation that causes so much pain.

If, in conclusion, we return to the debates about editing Clare, we can see how grounding ethical claims in rhetorical figures, in making Clare live and speak through the editorial process, would seem to register a different situation. But is this scene so very different? Aren't spelling, punctuation, and grammar the central issues in each of these debates? Isn't Clare made to live, speak, and be heard so that he can tell us something about his commas and periods? And if these commas and periods are more than mere marks, caesurae, and drillings, if they are the vehicle through which Clare becomes a poet, the vehicle furthermore of his living on and speaking to us and of our recognizing him, are these editors not personifying grammar at the very moment that they insist upon its sheer importance? Does the personification of Clare posit him not only as an ethical subject (the subject of justice), but equally a grammatical subject, to recall the last line of the other poem he wrote called "I Am," "that's all." And perhaps, then, it is no great surprise that for all their animosity, the difference between Bate's and Robinson's editions of the poem comes down to commas and dashes.[53]

Doing justice to Clare is a real temptation: Who wouldn't want to remedy the harm done to him through years of betrayed allegiances, indifferent audiences, and unrealized success. But justice, in particular a late justice ordered by an effaced prosopopoeia, can only fail to accomplish its aims: It becomes an aesthetic rather than an ethical project, or a moral rather than an aesthetic project, or merely a grammatical one. This is not a problem, per se, except that it relies upon the exclusion of this conditional possibility. The case of Clare, perhaps, should stand for us as a quiet warning as we recur to ethical terms and temptations in our own critical practices. What kind of justice can criticism accomplish? To whom or to what does it address itself? And is this addressee, rather than a voice we listen to and amplify, instead a voice we posit and figure? It may be that so long as critical claims take the form of ethical claims, they also will rely upon a figural or aesthetic act from which ethical acts may be indistinguishable. Perhaps this helps us to register a critical scene that is not structured by a dialectic of creation and salvation, one in which Clare's force, as Ashbery suggests, is like a "jump-start," that is, a possibility of mutual animation generated by a rhetorical figure. The example of Clare reveals that as editorial (and critical) interventions become more and more heavily (and at times dogmatically) focused on ethical claims, they also forget the very rhetorical gesture that is their condition of possibility, which is also, as I have tried to suggest, the condition of their impossibility.

4. Inventions of Self-Identity

In 1841, John Clare invents the term "self-identity." The context is an unlikely one: a fragmentary essay—or prose poem—in which he insists upon the moral necessity of self-recognition. Although Clare is virtually powerless in other domains, he discovers that he may be the only one capable of recognizing himself and of maintaining his place in the world. He thinks that self-recognition will ameliorate the risk of his disappearance or of his life becoming a form of social death. And so, at more or less the same moment that Francis Galton and Alphonse Bertillon develop biological profiling as a means of radical identification, Clare invents another form of identity, *self-identity*, by which he means not only the identity *of* a self, but identification *by* a self beyond recognition. As he writes:

> **Self Identity**
> A very good common place counsel is *Self Identity* to bid our own hearts not to forget our own selves and always to keep self in the first place lest all the world who always keeps us behind it should forget us all together—forget not thyself and the world will not forget thee—forget thyself and the world will willingly forget thee till thou art nothing but a living-dead man dwelling among shadows and falshood
>
> The mother may forget her child
> That dandled on her lap has been
> The bridegroom may forget the bride
> That he was wedded to yestreen
>
> But I cannot forget that I'm a man and it would be dishonest and unmanly in me to do so
>
> Self Identity is one of the finest principles in everybodys life and fills up the outline of honest truth in the decision of character a person who denies himself must either be a mad man or a coward

> I am often troubled at times to know that should the world have the impudence not to know me but willingly forget me wether any single individual would be honest enough to know me – such people would be usefull as the knocker to a door or the bell of a cryer to own the dead alive or the lost found there are two impossibilities which can never happen – I shall never be in three places at once nor ever change to a woman and that ought to be some comfort amid this moral or immoral 'changing' in life – truth has a bad herald when she is obliged to take lies for her trumpeters – surely every man has the liberty to know himself
>
> Tis Liberty alone that gives the flower
> Of fleeting life its lustre and perfume
> And we are weeds without it.[1]

At first glance, Clare's account of the self is remarkably contemporary, especially when it is understood to describe not only the precarity of individual psychology but also the possibility of political existence within and beyond recognition. One way to perceive the strange force of a self-identity that emerges apart from recognition, and indeed, without identity in a conventional sense, is by reading it alongside Giorgio Agamben's discussion of personal identity in modernity. In "Identity without the Person," Agamben tracks a significant change in the modern understanding of persons, which he considers as part of a new political preoccupation with life and survival, to argue that emerging technologies, above all biometric means of identification developed in the nineteenth century and that have become more sophisticated and all-encompassing in the twenty-first century, have changed what it means to be a person and what identity signifies. For Agamben, these technologies, above all fingerprinting in its old and new modes, imply that biopolitics not only reflects a conception of populations but also bears upon personal identity.

Agamben opens "Identity without the Person" by focusing on the etymology of *person* or *persona* and its originary meaning as a mask.[2] He acknowledges that personal identity, beginning with the Romans, depended upon a persona, or a mask, that functioned like a name and through which one found a place in social, juridical, and moral life. He goes on to explain that "*persona* came to signify the juridical capacity and political dignity of the free man. . . . The struggle for recognition is, therefore, the struggle for a mask, but this mask coincides with the 'personality' that society recognizes in every individual (or with the 'personage' that it makes of the individual with, at times, reticent connivance)."[3] "Persona" is a name for the condition of being a citizen, bearer of "juridical capacity," "political dignity," and freedom, all of which are the effect not of birth, but of bearing a mask, at once a gift of personality and specter of typification. Agamben continues by recognizing that the *persona*—the person or mask—has not only a juridical and social function but also a moral one, which here implies both *identification* with a social or public persona and *nonidentity* with it. The mask signifies the person, but the person understood as an excess or residue that outstrips the mask that is its condition of recognition and recognizability. Agamben suggests that with modernity the residue that reveals a gap between the man and the mask, like the gap between a political citizen and his biological life, erodes. In this case, the erosion occurs as an effect of technological forms of recognition that take the place of personal ones, through a new scene of recognition that occurs in the absence of others, and as the product of a range of identificatory machines.

Agamben perceives this change as intimately tied to forms of violence and exclusionary acts based upon biological data. The rapidly expanded appropriation of biological data has major consequences, he argues, for not only could it transform the capacities of a sophisticated genocidal regime, allowing for an extermination that could be "total and incredibly swift," but moving from individual lives to the concept of living in general, he offers the provocation that "even more serious . . . are the consequences that the processes of biometric and biological identification have on the constitution of the subject."[4] This new biological identity at once solves the problem of recognition and renders it a machinic rather than a human operation, exposing a form of total (and totalitarian) identification that is at odds with personal identity.

As a result, moral relations dependent upon the promise of a personality collapse, while virtual masks, both "artificial and illusory," proliferate on Instagram, Facebook, and Twitter.[5] Unlike the mask that in an earlier moment served as one's persona and personality, at once the index of responsibility and that for which one was responsible, these virtual faces (or *profiles*, as they are understood both in law and social media) are unaccompanied by a promise because, if Agamben is right, identity comes to rely not on social, but on merely technological relations.[6] This is what he calls "identity without the person"—identity without recognition, gesture, communication.

As much as Agamben abhors this condition, in which identity is predicated upon machines that scan our retinas rather than other beings who look into our eyes, he resists a pure nostalgia. Instead, he offers a messianic gesture, even as he claims to engage in a project that neither laments the bitter past nor anticipates a guaranteed future, proposing that "we must be prepared, with neither regret nor hope, to search—beyond both personal identity and identity without the person—for that new figure of the human."[7] Yet, at the moment that Agamben introduces this quest for a new figure, he also recognizes that he is in search of a figure that may no longer be human, and a face that is no longer identical with human existence. He is willing to forgo the human as the name of this condition and as the source of a new experience of identity and recognition, conceding that, "perhaps, what we must search for is simply the figure of the living being, for that face beyond the mask just as much as it is beyond the biometric *facies.*"[8]

Agamben searches for a face that does not stand *behind* the mask, but exists *beyond* it, as well as beyond biometrics. This is the sign of another order of recognition, or of an experience of others that occurs in the absence of recognition. As a result, this figure has the structure of an Event: it cannot be seen; it does not arrive. The idea of it, its possibility, confounds us, a bit like Clare's Mary. It interrupts both wakefulness and dream, for it suggests the possibility of a radically new experience of being with others, a movement from "the persona" beyond "identity without the person" to "the face beyond the mask." Thus, what is at stake in thinking about identity is not only thinking about the self but also the exposure of the rhetorical figure as a condition of living (or life) in which survival and social relation no longer depend upon recognition. At first glance, this figure of life without recognition appears as a figure without identity. Yet, however new, its history is also a long one, for nonvisibility has allowed this figure to exist in an ongoing, permanent state of privation and, recalling Derrida, *survivance*. Indeed, this new figure is also figure itself, the figure that has invisibly accompanied the invention of the lyric subject, the drama of recognition, and the exhaustion of subjectivity through total identification. While Agamben calls his essay "Identity without the Person," naming the condition of modern biopolitical existence, the life that he anticipates is a form of survival without recognition that solicits interaction without relation and relies upon presentiment rather than sense.

Clare's "Self Identity" arrives just on the cusp of this transition from personal to biological conceptions of identity. Clare wrote "Self Identity" fifty years before Galton outlined his theory of fingerprinting, and less than thirty years before the first edition of his *Hereditary Genius* appeared, a book that brings scientific methods to bear upon the popular notions of inheritance and talent that came to define Clare's place within the literary sphere, and that Evelyn Fox Keller identifies as the first time that nature and nurture are conceived in opposition to one another. When Clare writes "Self Identity," he is between two asylums and obsessed with recognition as a condition of existence. He argues that self-recognition, or self-identity, is a way of avoiding a life relegated to nonrecognition, or what he considers living death. Self-identity, in this sense of self-recognition, is a way of managing others' failure to recognize the self and of maintaining social viability in the face of perceived (and real) abandonment. For Clare, it is not merely a matter of remaining beholden to one's mask as a moral obligation (as Agamben holds is the case for the Romans), but rather of grasping a living being, a self, in the face of its disappearance or nonvisibility. Clare insists that one must remember what others forget, hold fast to what they leave behind. In this sense, the self maintains its position not only in the act of self-positing (of being one who can recognize or remember), but also in self-recognition.

For Hegel, with whom Agamben opens "Identity without the Person," recognition allows one to live as a person. As Agamben recalls it: "The desire to be recognized by others is inseparable from being human. Indeed, such recognition is so essential that according to Hegel, everyone is ready to put his or her own life in jeopardy in order to obtain it. This is not merely a question of satisfaction or self-love; rather it is only through recognition by others that man can constitute himself as a person."[9] Clare, too, perceives the necessity of recognition, which he understands in terms of remembering and knowing. Yet he remains unconvinced that recognition *must* depend upon others, or even that it *can* depend upon them, and he is willing to take things into his own hands as needed. More than this, true recognition of an other, recognition of one who cannot be recognized, because his speech or work belongs to no preexisting genre, because he cannot comfortably live in society or endure exile to the asylum, because his place in the world is marked by a conditional displacement, demands not conventional forms of sense and sensibility, but rather a capacity for recognition that occurs without knowledge and in the face of preexisting frameworks that would consign such recognition to impossibility. In other words, the form of recognition that Clare demands, a recognition that ultimately leads to his dependence upon (and invention of) self-identity or identification as the work of a self, suggests that self-identity is a form of relation that occurs in the face of impossible recognition in which "what I am none cares or knows."

Clare explains self-identity as an act of self and one that is a condition of personal existence and social relation. As he writes: "A very good common place counsel is *Self Identity* to bid our own hearts not to forget our own selves and always to keep self in the first place lest all the world who always keeps us behind it should not forget us all together—forget not thyself and the world will not forget thee."[10] Clare's gesture does not merely place autonomy in opposition to community, but registers autonomy as the precondition of recognition by others. To use Hegel and Agamben's notions, it becomes the precondition of one's "constitution as a person." Indeed, as Clare continues, in the absence of self-recognition, one risks disappearance and living death: "forget thyself and the world will willingly forget thee till thou art nothing but a living-dead man dwelling among shadows and falsehood."[11] For Clare, this act is a moral responsibility, and although he often complains about how others treat him ("my friends forsake me like a memory lost" or "Neglect is the rust of life that eateth away and layeth the best of minds fallow and maketh them desert"), here he suggests that recognition is an autonomous act and personal responsibility. One's existence as a person is not initially dependent upon others but upon the self.[12] Thus, self-identity in Clare's account is not a static state, but a process of recovery and recognition that responds to and protects against social abandonment.

When Clare insists upon the necessity of self-identity as a means to survival, he does so with the understanding that the world (as other men) has the power to consign him (or any one of us) to life or living death. He holds that prioritizing the self will save one from depending upon the taste of others and that self-identity is an act of fraught survival that occurs beyond or despite the limits of recognition. For Clare, this is nevertheless also a normative position, one that assumes rationality ("good common place counsel"), as well as masculinity and honesty ("I cannot forget that I'm a man and it would be dishonest and unmanly in me to do so").[13] As such, Clare can be understood not only to respond to a situation of absent sensibility, but also in doing so to wrest from the world the power of recognition (the power over his existence as a person, or in his terms, the power to turn life into a form of death). He posits a form of self-responsibility that approaches self-sovereignty rather than mutual interdependence, and yet, this sovereignty produces (and is conditioned upon) another scene self-division.

Responsibility for Clare is not, as it is for Emmanuel Levinas or even Judith Butler, dependent upon the *refusal* of the violence of subjectivity, instead he relies upon self-recognition in order to counter the violence of a persecutory world. At the same time, self-recognition is a response to handling the impossibility of recognition by others, a way of living with the experience of one's own nonidentification and evasion of sense. Rather than breaking up the structure of sovereign violence, Clare appropriates it. And yet, in doing so, he also calls our attention to a self that is otherwise unrecognizable. This is a self who suffers from the experience of multiple identities; who has been taken as others' figure—a peasant poet and a mad poet, as a figure of working class and ecological poetics, as a patient and as a genius; and whose existence is an experience of permanent displacement, which is to say one who lives between projection and spectrality.

In order to illustrate the urgency of self-identity in the face of potential abandonment and misrecognition, Clare turns to Robert Burns's "Lament for James, Earl of Glencairn." The poem is a hyperbolic tribute that hails

Glencairn, Burns's benefactor as unforgettable and admits that even strong guarantees against abandonment, social formations like the family or the couple, are weaker than his devotion. The poem concludes:

The bridegroom may forget the bride
Was made his wedded wife yestereen
The monarch may forget the crown
That on his head an hour has been
The mother may forget the child
That smiles sae sweetly on her knee
But I'll remember thee Glencairn
And a' that thou hast done for me!

Yet, when Clare insists upon his devotion not to another, but to the self as a masculine subject, he reworks Burns's account of fragile institutions not in order to demonstrate the strength of his memory (as Burns does), but to elaborate the fragility of his position:

The mother may forget her child
That dandled on her lap has been
The bridegroom may forget the bride
That he was wedded to yestreen

But I cannot forget that I'm a man and it would be dishonest and unmanly in me to do so.[14]

In citing Burns's poem, Clare remembers (even if he inverts) the examples in which an excessive sense of devotion reveals the weakness of social relations and the possibility that the unforgettable could be forgotten. But Clare elides one example of this forgetting: that of a man—the sovereign—who forgets himself and his position.[15] This elision in which one forgets who one is, in which the king forgets his place and power, is of another order. In some sense Clare *must* exclude this possibility as an example of fragility because his entire essay is organized around establishing the impossibility of *self*-forgetting, whether as fact or charge. In Burns's poem, it is the dead benefactor, rather than the living poet who remains unforgettable. While Burns's actual relationship to Glencairn is reported to have been somewhat more ambivalent than this poem lets on, and the poem's wild hyperbole could be read as an ironic sense of commitment, the poem nevertheless memorializes a scene of external recognition: "a' that *thou* has done for me!"

Clare, too, was not without his own benefactors, including John Taylor,

Eliza Emmerson, Lord Radstock, and Earl Fitzwilliam, as well as those like William Knight, who preserved his manuscripts, and Frederick Martin, who wrote his first biography. But at this moment, Clare understands them only as members of the world that has failed to remember him. Where Burns acknowledges a debt in his poem, Clare affirms only his own solitude and the demand that he must sustain himself. He offers not a positive statement of promised memory in the face of forgetting, and with that promise a future (Burns: "I'll remember thee Glencairn"), but rather a negative statement of forced vigilance and a fixed identity conceived not even as personal identity, but rather as gender identity, which here is akin to species identity. In other words, rather than remember, Clare states that he cannot forget; and rather than hold onto an individual, Clare manifests himself as belonging to a species, as having personhood. As he concludes: "I cannot forget that I'm a man and it would be dishonest and unmanly in me to do so."

At least two things occur in this rewriting. While Clare excises Burns's example of sovereign self-forgetting (the monarch who forgets the crown), he also reintroduces it in another form when he turns self-identity into impossible forgetting. However, the self here is merely the possibility of having an identity; it is not the memory of the individual self, but of mere belonging. In Clare's rendition of the poem, others might forget him (like the abandoned women and babies that appear here), but he can never forget that he is a man. Moreover, while Burns identifies a moment in which the living remember the dead, Clare places *himself* in the position that Glencairn holds, placing *himself* in the position of the dead who will be remembered, and of the benefactor who sustained the poet and to whom the poet remains indebted. Clare does this as a charge. In other words, when Clare writes, "But I cannot forget that I'm a man," he replaces the final lines of Burns's lament for another with an obligation to remember the self as a rememberable human being. Inhabiting the place of both the living (Burns) and the dead (Glencairn), both the poet who laments and the benefactor who has passed, Clare also inhabits the place of both the one who recognizes and who is recognized. But at the moment that Clare turns a scene of mutual recognition into one of solitary self-grasping, at the moment that he remembers that he is a man, he also inhabits the position of a dead man, albeit one who Burns promises never will be forgotten. If here Clare struggles to avoid a life of social abandonment and living death by taking his life into his own hands and rendering the self responsible for its own recognition, the very gesture through which he claims his life as unforgettable puts him in the place of the dead. This life is that of a survival that occurs in the place of the dead, and a new "figure of the living being" whose condition exceeds sense and visibility and who lives beyond recognition. This occurs at the very moment that the term "Self Identity" first enters into the English language.

In some sense, Clare's "Self Identity" anticipates the situation that occurs in his later poem, "I Am," which opens with an apparently unequivocal assertion of self-presence, yet one for which identity is withheld: The *I* is without content. Despite appearances, the poem states something radically other than the speaker's self-identity; it opens not with "I Am I" or with any positive or unified statement of self. Rather, before arriving at any content, it turns on the withdrawal of interest ("but what I am none cares"), and the dependence upon the perception of the self becomes (again despite appearances) a statement of radical nonknowledge ("or knows"). I already have considered how inscription and punctuation interrupt the working of figure in the poem and more generally in the reading and editing, which is also to say, in the recognition, of Clare, and I have suggested that "I Am" emerges as a poem of "depersonification" or nonpersonification. While "Self Identity" is ordered around the task of remaining a man and fostering a life that is not a form of death, a task that is fulfilled through vigilant and unfailing acts of self-recollection, when read in the shadow of "I Am," this essay seems to advocate not for a person, a self with qualities, but rather for the self as a capacity and obligation for self-reference, self-recollection, and self-recognition in the absence of sensibility. In other words, this position could be perceived as a purely formal one. Yet, if the self is merely a formal position (permanent self-reference) and if the recognition advocated here is that of pure self-preservation, what is preserved is a position rather than a person or put another way, what is preserved is a person or identity in the absence of sense, one that, as the rewriting of Burns suggests, displaces the sovereign and recovers it in the place of the dead.

Indeed, for Clare self-identity is not personality, and when it comes to personalities, Clare admits several. While here Clare declares his static physical and biological identity, proclaiming, for example, "I shall never be in three places at once nor ever change to a woman" and while he raises the specter of madness when he exclaims that "a person who denies himself must either be a madman or a coward," he only indirectly refers to his own experience of self-*forgetting* and the "affective disorder," to use Jonathan Bate's term that led him "to believe that he was Lord Byron—or Lord Nelson, or Jack Randall the boxer."[16] In some sense, the entire essay emerges as a performance of nonself-identity even as it advocates for self-identification and bodily immutability. Not only does the essay forget or rewrite the very condition of its emergence, turning from psychic or poetic identity to the self as a biological and physical form, but it also anticipates the condition that Agamben calls "identity without the person," albeit from a position that claims biological absorption as a response to personal insentience.

Thus, it is perhaps no surprise that Clare comes to "invent," as Jonathan Bate has argued, not only the phrase "self-identity," but also "nonself-identity," which appears in "Invite to Eternity," a late poem that describes the dissolution of the world of the subject and the instability of physical forms.[17] In that poem, Clare encounters a world of radical formlessness, yet one in which social relations endure nevertheless. In fact the poem, an *invitation*, is organized around an address to a woman who the lyric subject asks to accompany him "Through the valley-depths of shade, / Of night and dark obscurity" (ll. 3–4). The poem, made up of four stanzas, offers a series of negative moments and scenes of dissolution in which even the examples are not discrete, since they are connected by genitives, anaphoras, and parataxes that blur their boundaries. Yet, there is one thing that remains unchanged in this poem: the subject and, potentially, his beloved companion, for their destiny is beyond temporality.

An Invite to Eternity
Say, wilt thou go with me, sweet maid,
Say, maiden, wilt thou go with me
Through the valley-depths of shade,
Of night and dark obscurity,
Where the path hath lost its way,
Where the sun forgets the day,
Where there's nor light nor life to see,
Sweet maiden, wilt thou go with me?

Where stones will turn to flooding streams,
Where plains will rise like oceaned waves,
Where life will fade like visioned dreams
And mountains darken into caves,
Say, maiden, wilt thou go with me
Through this sad non-identity,
Where parents live and are forgot,
And sisters live and know us not?

Say, maiden, wilt thou go with me
In this strange death of life to be,
To live in death and be the same,
Without this life or home or name,
At once to be and not to be,

That was and is not—yet to see
Things pass like shadows, and the sky
Above, below, around us lie?

The land of shadows wilt thou trace
And look—nor know each other's face,
The present mixed with reason gone,
And past and present all as one?
Say, maiden, can thy life be led
To join the living with the dead?
Then trace thy footsteps on with me:
We are wed to one eternity.[18]

The poem draws upon the language of both "I Am" and "Self Identity." It describes a possibility of relation in the absence of recognition, in a time abandoned by the sun, in a place in which physical property is turned radically unstable—stones become streams, plains become waves, "where there's nor life nor light to see." Here it is not that blindness is a property of the self, but rather the conditions under which vision, recognition, or identification could be possible do not exist. Once again, even family relations disappear despite the ongoing lives of their members. And the face of even one's most intimate companion is unknown. This forgetting, which evokes the abandoned women and children in "Self Identity," is multidirectional: It occurs both across and within generations, by others and by the self. Yet, the invitation describes not an evasion of this world (as death), but rather a dive into it; it describes not radical loneliness and abandonment, but, despite calling the scene one of "sad non-identity," it figures a relation both potential (ordered by an invitation) and actual (effected by the invitation) that exists beyond temporal limits. Here there is no anxiety about self-identity, only the surety that identity endures "in this strange death of life" and even when we "live in death."

Put another way, self-identity names survival within difference and privation; "sad non-identity" names the time in which one lives on beyond temporalization and the opposition between identity and nonidentity, where one lives "without this life or home or name." This eternity, which is not a theological eternity, but rather the name for a shadow world defined by radical indistinction and shapelessness, emerges as a place: the poem's destination. It is not simply a relation between two, but a relation to another relation to living, and one that has the shape of a marriage. It takes on this marital form, which, is not between two, but *of* two to a possibility of living. It is an act of unification that depends upon an unexperienced possibility (marriage) that is displaced onto a past that never will have taken place. While each of the first three stanzas concludes with a version of the invitation and an implicit question mark (which Bate inscribes as an actual one), the final stanza is an announcement that assumes the fulfillment of the invitation; it assumes an affirmative response that never will have been uttered and that leads to a permanent present (and presence) of being "wed to one eternity." It produces marriage as a consignment to a future in which the difference between the living and the dead, past and present is suspended.[19]

In Clare's essay on self-identity, an insisted upon, morally urgent identity emerges as pure self-referentiality and recurring self-recognition; it appears to be a structural or performative position, rather than a substantial one.[20] This position can explain how recognition occurs beyond sense, further fleshing out the shape (or shapelessness) of a self whose survival coincides with nonsense. It is at once abstract and functional, and does not become the basis of a kind of transcendence as one might expect in conventional accounts of romanticism. Rather, it is a ground zero or baseline to which all performances, delusions, aspirations, and relations revert.[21] Self-identity does not allow one to get beyond the self, or even get to the self. It ensures that there is something—or something *and* nothing (what in "An Invite to Eternity" Clare calls "At once to be and not to be")—in lieu of nothing, even if this something remains a process of mere self-referentiality rather than complete self-awareness or self-consciousness, even if it remains beyond sense. This also explains why in "An Invite to Eternity" Clare can describe the resilience of life in death ("to live in death and be the same / Without this life or home or name"), for being the same does not rely upon character or personality or biological data, neither life, nor home, nor name. This is an existence in which being and nonbeing are nonopposed in the present, an existence whose possibility, in other words, exists in and through an act—that of marriage—that is the moment at which a poetic utterance, an utterance that remains nonreferential and enigmatic, comes to have stakes in and for the world. This marriage beyond recognition and recognizability, between those who do not even "know each others face," suggests a possibility of living—survival—beyond recognition and temporality.

In *Vita: Life in a Zone of Social Abandonment*, João Biehl asks "What kind of subjectivity is possible when one is no longer marked by the dynamics of recognition or by temporality?"[22] If Clare's "An Invite" offers one response, Biehl takes up the question through another experience of poetry and survival. For Biehl, the question and the answer both emerge in an ethnographic account, at once devastating and hopeful, of Catarina Gomes Morales, a woman abandoned by her family and left to live—or die—and to write in an asylum called Vita. Vita exists at the very edges of institutional existence in Brazil and, as Biehl explains, Vita "creates a humanity caught between visibility and invisibility and between life and death."[23] Biehl calls this posthumous condition that of the "ex-human," and he goes about tracking, through the case of Catarina, how a person can become other than a person, how one might pass from existence within a family and home to a life of abandonment.[24]

Biehl's question about the possibility of subjectivity in the absence of recognition and temporality resonates with Agamben's inquiry into the possibility of a new figure that exists beyond previous modes of recognition and Clare's account of an existence that endures despite the indifference of others. Yet, whereas Agamben turns toward culturally inflected philosophy, and Clare takes up autobiographically inflected poetry, Biehl uses the ethnographic and ethical approach of anthropology. Biehl's question about the possibility of subjectivity may be a rhetorical one, suggesting that no subjectivity is possible, as his constant failures to help Catarina recover a place within social and institutional existence demonstrate. But, he also is unwilling merely to consign his subject to a position that Agamben has called bare, or naked, life. As he explains, "A human form of life that is no longer worth living is not just bare life—language and desire continue." Now, as much as I understand what Biehl means when he undertakes to distinguish language and desire from bare life, and implicitly to distinguish the human and the animal, language is not merely absent from so-called bare life, but endures without being the marker of human existence. In other words, Catarina's language, which demands a new dictionary, like Clare's vernacular and Paul Celan's radical testimony reveals another relation between language and life.[25]

For Biehl, this form of living also requires a new method, an ethnography oriented by forms of listening that evoke literary reading and psychoanalysis. He explains: "As I listened to and excavated what had made Catarina's voice 'posthumous,' a life force—often gaining form in the figure of the animal and related to libido, belonging, and opposition to a death drive—emerged to rework thought, social relations, and family life. Ethnography became the missing nexus between the real of Catarina's body and the imaginary of its mental and relational schemes, between the abandoned and the family, the house and the city, individuals and populations in Vita."[26] Ethnography, in this case, is the ethnography of an individual subject. It aspires to rearticulate the broken links to home, family, legibility, and world that Catarina suffered. Through this form, Biehl undertakes to bear witness to a posthumous existence and transform it into a life that makes sense and whose symptoms and sufferings belong to networks that can be identified and reimagined. Indeed, ethnography is a means "to own the dead alive or the lost found," to recall Clare's account of recognition by others. It is not a matter of a powerful sovereignty, but rather the recovery of a life otherwise lost or abandoned for death.

Biehl first discovers Catarina in 1995. The volunteers who worked in Vita explained to him that "they knew nothing about her life outside . . . they said that she spoke nonsense that she was mad (*louca*). She was a person apparently lacking common sense; her voice was annulled by psychiatric diagnosis."[27] Biehl goes on to recover her name, her story, and in some sense her sanity or at least the meaning of her condition. In other words, he rescues her among the others like her left to wait for or, as he puts it, "with" death, and transforms her anonymity into a name and what is left of her life into a story. She becomes someone who could be known and who can live. Part of Biehl's fascination with Catarina stems from her interest in language and writing. While she is wasting away, alternately hypermedicated and undermedicated, in Vita, the ironically named asylum where she is surrounded by drug addicts, the extremely poor, and others, like her, with illnesses that remain unnamed, undiagnosed, and untreated, Catarina, by her own account, is at work to write a "dictionary."

The dictionary, a compendium of words that archive her experience and condition, is a form of poetry; at least this is how Biehl imagines it when he divides the lists of words and phrases and suggests that "we might face Catarina's writing in the same way we face poetry." Her project, as she tells Biehl, is one of remembrance, as he explains, "She was doing this 'to not forget the words.'" He adds that she produced "strings of words containing references to persons, places, institutions, diseases, things, and dispositions that seemed so imaginatively connected that at times I thought this was poetry." Like William Knight, who preserved Clare's asylum poetry, Biehl saves nineteen of the twenty-one notebooks that Catarina filled with words, and like John Taylor, he "translated some of them and . . . placed them in the form of stanzas." Some of Catarina's lines register her anxieties about survival, including the first Book that Biehl includes, which opens: "I offer you my life / Dead alive / dead outside / alive inside." Others deal more directly with questions of the self: "I, who am where I go, am who am so." Biehl's interest in Catarina is part of a larger effort to understand what he calls "life in a zone of social abandonment," and the acknowledgment, however painful, that "there are places in the present, even in a state founded on the premise of inviolable human rights, where these rights no longer exist, where the living subjects of marginal institutions are constituted as something other, between life and death."

Over time, Biehl's interest becomes not only ethical or academic but leads him to know Catarina and to allow her to become knowable. This shift reveals the role of biology (and diagnosis) for recognition by both the family and the state. In other words, although Biehl asks us to consider the place of subjectivity beyond recognition and temporality, his project is equally one of alleviating the suffering that attends this absence, and he discovers that in order to do so he must find new ways of caring for those who have been disavowed. His project ties ethnography to an effort at responsiveness, reflecting the responsibility an anthropologist faces with not only appropriating another's story but also intervening in and striving to improve it. Biehl's work tracks how ethnographic biologization becomes a means through which mysterious symptoms can come to be diagnosed and as such rendered treatable (or untreatable), and how Catarina can become recognizable to others and returned to her family history if not to her actual home once the symptoms that prompted her family to view her as an alien madwoman and abandon her to Vita come to be understood as indicators of a rare genetic disease that they all share. Yet, there remains one risk here, and that concerns the place of science (social or biomedical) and its role in conditioning recognizability. Indeed, while Biehl goes to extraordinary lengths to rescue Catarina from her destiny in Vita, efforts that despite their intensity cannot rescue her from death, this work nevertheless relies upon the production of biological claims to citizenship and the framing of Catarina's experience as one that is recognizable to biomedicine.

In other words, recognizability remains the condition of care, even for a reader who refuses conventional oppositions between science and poetry. As in "I Am," knowledge and care continue to be soldered together rather than disarticulated as the opening of a new possibility of attention that does not rely upon recognition. Biehl, who becomes intensely involved in Catarina's project, locates her medical records, introduces her to doctors, and informs her family members of their condition. At the same time as he salvages Catarina as a human being, that is, at the same time as he undertakes to care for her, feeding her properly, seeking out medical care (beyond the random allocation of drugs that occurs in Vita) and perceiving her as a poet, not merely a madwoman who struggles with her words, he also affirms, rather than challenges the conventions of biological citizenship. His project, which he describes as a detective story, entails the transformation of Catarina into a patient and a subject (anthropological, poetic, human), rather than one of the living dead "dwelling among shadows and falsehood." Yet, the question remains: Can she be a subject of care in the absence of biologization, its temporalities and its conception of recognition? Can her words matter if they are not made into stanzas and turned into poems? The challenge may be too abstract for a case like Catarina's, which required a certain degree of urgent responsiveness, as it asks whether we are ready for a new figure of living being, ready for forms of relation that are not dependent upon recognition, or whether we are ready for a figure that is not new per se, but ever imperceptible as it depends upon another relation to recognition and to care? This is the shape of a subject who could be an object of care without knowledge or recognition. In other words, this is the subject beyond shape, an Event, like the marriage to eternity that Clare describes in "An Invite," that binds without ever belonging to a present.

While I have shown how Clare and Catarina might form a constellation for thinking forms of identity and recognition, and while I have suggested that taken together we might be able to think with them a relation between life, social abandonment, and poetry, there also are significant differences between Clare's experience and that of Catarina, and these differences interfere with a too complete conflation of their conditions. In the first place, while it is true that our access to Clare is mediated by the editions of his poetry, the manuscripts that Knight and others have saved and the biographies that narrate his life, we nevertheless remain in a position to compare manuscripts, read extensively in the writing published in his lifetime and posthumously, and access texts that precede his asylum period. We cannot say the same for Catarina, whom we read only through Biehl's ethnography. She is a witness to the limits of subjectivity and recognition, brought into being through another's testimony.[28] This difference suggests that our reading is mediated not only by the norms of literary taste or editorial style but also by the norms of universities and knowledge production in the human sciences as well as contemporary protocols that are both legal and disciplinary. The kind of subjectivity that Biehl discovers through Catarina has a structural difference from the inquiry into the self that Clare undertakes 150 years earlier. Moreover, while it is tempting to draw a line between the biologization of subjectivity and its displacement, and to assume a clear continuity between English asylums of the nineteenth century, where Clare came to write "To Mary," "I Am," and "Invite to Eternity," and zones of abandonment in the twentieth-century, like Vita, where Catarina writes her dictionary, the story is also more complex.

At about the time that he writes the essay on self-identity, Clare had become, as Parson Mossop reports, using more or less the same formulation used for Catarina 150 years later, "a stranger to his own family."[29] Mossop's response (and that of Clare's family) is to return Clare to the asylum, this time to the Northampton General Lunatic Asylum, where he would remain for the rest of his life—never again to see his wife, Patty. If Clare is unrecognized and unrecognizable by his family and friends (the very forgetting that, via Burns, he reflects in the essay on self-identity, and that we encounter, years later, in the example of Catarina Morales), he is also unrecognized by his doctors, according to Roy Porter. While there has been much critical (and uncritical) speculation about the nature of Clare's mental illness, Porter finds it curious that his doctors "had pathetically little illuminating to say; and some of their statements appear palpably false or, at least, highly contentious."[30]

Porter offers three explanations for the paucity of insight into Clare's condition: the first concerns the fact that Clare's time in the asylum was good for everyone, and diagnosing and then curing his illness might have interfered with the more general benefits of the asylum on his well-being and that of his family; the second concerns the state of Victorian psychiatric practice; and the third concerns the popular conception of biological or organic sources of mental illness. More specifically, Porter suggests that the asylum was a way of managing rather than curing Clare. He was able to live there in relative health without having to labor in the fields. Staying in the asylum solved a problem of Clare's livelihood The labor to which he would have inevitably returned upon release would have rendered his life a living toward death. Porter suggests further that Clare's placement was a boon for the hospital, which now had a poet in residence, and also a salve for the literary public who could receive his poetry and not worry about his survival. Moreover, prior to Freud, Porter reports, "doctors deliberately prevented patients from talking about themselves, lest this encourage delusions of grandeur, morbid introspection, over-excitation or egoistical hysteria."[31] Clare's worries about not being recognized therefore might not have been purely paranoid. Instead they can be understood as symptoms of his displacement within the home, where he had become a stranger, and of the asylum's conventional practice of discouraging speech. Finally, Porter suggests that it is the very biologization of mental illness that precluded "intense or intimate *discursive* encounters with patients."[32] If Porter is right, assumptions about the organic nature of illness interrupt rather than facilitate recognition. This is not (or not yet) a matter of recognizing categories of illness and cure, but of understanding biology, rather than the unconscious, as the source of mental disturbance.[33]

Both John Clare and Catarina Morales expose the question of how or whether an individual who falls outside of the conventions of visibility can survive and how or whether they can continue to be objects of care. Biehl's *Vita* suggests that using the tools of ethnography, anthropology, even a sort of archaeology, one can and must rescue an abandoned person by returning her to the categories of social recognition, diagnosis, kinship, and human rights. Yet, Clare offers a different challenge, one that may be more radical, even if I in no way want to underestimate the extraordinary value and generosity of Biehl's engagement with Catarina and his willingness to enter into a world that is nothing less than hell on earth. Clare challenges us, despite what he says he wants, to stay on the side of nonvisibility. He invites us to this eternity, which may be, as in "I Am," utterly catastrophic, compelling us not only to marshal the tools of biological citizenship but also to resist the conflation of visibility (or recognition) and care. While this might appear to be just a poetic—or hyperbolic—possibility, one that seems weak in the face of actual demands upon human rights and scientific (social or biological) responses to privation, it could be that a poetic stance is precisely what science needs.

In this sense, romantic or lyric subjectivity emerges not as a myth of unity or sovereign power, but as a possibility of survival beyond recognition and experience, a possibility that Paul de Man evokes in his early essay on "Criticism and Crisis," where he tracks "the transition from anthropology to the field of language and, finally, of literature."[34] In the essay, de Man undertakes to show how literature (above all romantic literature) interminably names a form of consciousness that depends neither upon another subject nor an object, but merely exists as the "consciousness of the presence of nothingness." The form of consciousness, or subjectivity, de Man argues, coincides with literary language even at the moments when critics would see it as most deluded about what literature can be or do. What the subject knows is not anything at all, apart from the fact that there is nothingness, and he proposes that a truly radical science of man would obtain its rigor not merely by dismissing the subject (as, he suggests, structuralist linguistics or anthropology propose to do), but rather by recognizing what literature knows about the subject, that it is merely consciousness without object or consciousness whose object is nothingness. Thus, de Man concludes with a bold vision of social science and its relation to literature: "From this point on, a philosophical anthropology would be inconceivable without the consideration of literature as a primary source of knowledge."[35]

In other words, the evacuation of the subject in the name of science ends up renewing the question of the self, and in doing so returning to the form of knowledge to which literature has a privileged access. To the extent that social science imagines other forms of knowledge about the world or the subject, it remains deceived by the very discourse that it is tasked with demystifying. Before arriving at this polemical position, de Man suggests that when critical discourse dismisses the lyric subject, it lingers on the side of blindness, rather than insight. He explains:

> Here the consciousness does not result from the absence of something, but consists of the presence of a nothingness. Poetic language names this void with ever-renewed understanding and, like Rousseau's longing, it never tires of naming it again. This persistent naming is what we call literature. In the same manner that the poetic lyric originates in moments of tranquility, in the absence of actual emotions, and then proceeds to invent fictional emotions to create the illusion of the reality of others. But the fiction is not a myth, for it knows and names itself as fiction. It is not demystification, it is demystified from the start. When modern critics think they are demystifying literature, they are in fact being demystified by it; but since this necessarily occurs in the form of a crisis, they are blind to what takes place within themselves. At the moment that they claim to do away with literature, literature is everywhere; what they call anthropology, linguistics, psychoanalysis is nothing but literature reappearing, like the Hydra's head in the very spot where it had supposedly been suppressed. The human mind will go through amazing feats of distortion to avoid facing "the nothingness of human matters." In order not to see that the failure lies in the nature of things, one chooses to locate it in the individual "romantic" subject, and thus retreats behind a historical scheme, which, apocalyptic as it may sound, is basically reassuring and bland.[36]

To claim to do away with the subject, to dismiss it as naïve or romantic, rather than as the marker of a void for which it remains responsible, is to miss literature's part in the demystification of subjectivity. Thus, when de Man insists upon what could be called *the value* of the poetic subject, he refuses to see its life as other than a form of death; the positing of a figure (the subject) is here the index of a void. When scientists or critics misrecognize romanticism (or literature) as unaware of or covering over this void and respond by evacuating the subject in the name of science, they remain beholden to a myth of overcoming in lieu of a recognition of the failure to overcome. Social science—and all the more so in the years since de Man wrote the initial version of this essay in 1967—assumes that one can get beyond fiction or subjectivity by doing away with the subject.[37] For de Man, literature exposes that beyond fiction, including the fiction of the subject, there is merely nothingness, rather than data or insight. He suggests that the only way of saying something about this nothingness is through a fictional act whose aim is not to "fill in the void" transforming data into knowledge, but to "assert itself as pure nothingness, *our* nothingness stated and restated by a subject that is the agent of its own instability."[38] For this reason, literature is not merely the object of criticism, but a more radical experience of criticism; it is what exposes and understands criticism's object.[39]

In a move that apparently differs from his later approach to blindness and insight, de Man explains that romanticism or literature is the awareness of the subject as a self-projection rather than the naïve production of a unity of consciousness and self, as ironically coincides with the evacuation of the subject in supposedly mythical discourses of social science or criticism.[40] When the subject endures in a lyric or literary mode, this endurance renders literature the space where nothingness or antimyth is guarded. The romantic subject, for de Man is a form of self-awareness in a negative mode: this self-awareness is the awareness of nothingness through fictional means.[41]

If literature then takes the place of criticism or the social sciences (at least the philosophical social sciences of linguistics, psychoanalysis, and anthropology) in producing knowledge, what kind of knowledge does it produce? If the knowledge produced by literature is that of nothingness or nonunderstanding, if in this early essay de Man understands literature as the awareness of nonawareness, might literary demystification also serve another purpose, one subtly suggested by de Man, of ordering one's relation to this kind of knowledge? De Man understands literature as affording a more critical and comprehensive understanding of the science of man than the human—or social—sciences, this is because it perceives of the subject as nothing or an act of literature of which only literature is aware. For de Man awareness of the self as contentless, fragmentary, or fictional becomes the condition of the forms of experience assumed to be the domain of the objective social sciences. Taken together, one can say that science is most objective and least violent when it is most literary, not when it merely abandons the subject nor when it maintains it, but when it imagines the subject as a productive fiction, a kind of anthropomorphism, with the ability to unsettle the forces of knowledge and power. Put another way, de Man seems to retain a place for the subject, albeit in this negative mode, as a source of accountability (Butler) or demystification; the subject is conceived as the awareness of nonself-existence or the existence of a one "who can never fully be accounted for," but remains accountable.[42] Yet, the challenge of each of these conceptions of the empty self is to couple a rigorous demystification of self-identity with an awareness of the actuality and usefulness of its fiction. While de Man anticipates literature as the source of anthropological knowledge and while Butler exposes a possibility of responsibility severed from self-transparency or self-knowledge, there remains a question of literature and responsibility still to be fleshed out. Taken together, both suggest that a critical account of the self allows for knowledge of nonknowledge, or nonaccountability at the origins of accountability, yet these remain definitional moments, moments in which the subject is defined or redefined.

Here we might recall the moment at which Agamben undertakes to think a new form of identity that he suggests might occur with neither nostalgia nor hope, just as we might recall Biehl's inquiry into the possibility of a subject beyond recognition and Clare's production of an encounter with a subject beyond form and sense, whether Mary or in the "I Am" poems, the "I" itself. Could it be that what is at stake there is a way of "facing 'the nothingness of human matter'"? While Agamben suggests that this is a more or less historical scheme that leads from classical theater to modern biometrics to a still unnamed future, might we also take up de Man's proposal that we abandon a mythic view of self-identity and historic knowledge, that is, abandon the myth of the evacuation of the subject in the name of science and of the possibility of a form that is not a reiteration of nothingness, and instead begin to face this nothingness? Is this the coming future that Agamben anticipates? If it is, what might our relation to this being look like? Will it look like Biehl's ethnography of Catarina, like Clare's acts of self-recovery, or will it occasion another relation to the living? Is this coming being, this new form of life, a life of poetry?

If the task of poetry is to separate knowledge and care and to nurture a sense of care and interest that does not rely upon knowledge, even literary or negative knowledge, it is because poetry is an Event, and as such depends upon not being recognizable or understandable in advance. This is one way of understanding what Barbara Johnson sees as the political dimension of poetry and its figures. In other words, it is poetry that achieves—and names—the kind of figure that Agamben anticipates at the end of "Identity without the Person," just as it is poetry that offers training or experience for a kind of relation that operates outside of recognition and recognizability. It may be that the profound sadness that we encounter in Clare is tied to his inability to disarticulate ignorance and responsibility, specifically, self-ignorance and self-responsibility, a disarticulation that might have allowed him to live unencumbered by the immense suffering that left him in the asylum and a disarticulation, moreover, that he also trains us to see. In other words, trapped between blindness and insight (like Biehl, who also finds Catarina and rescues her, even if he can only do so through an expansion of the categories of legibility and biology, rather than a more thoroughgoing rethinking of the means of recognition itself), Clare reveals a possibility of living on that he cannot acknowledge and a form of life that he experiences, albeit in the negative mode of nonawareness and rejection. What we have is a turn from the poetics of life to a life of poetry.

5. The Poetics of Homelessness

In "Relocating John Clare," their introduction to the essay collection *John Clare in Context*, Hugh Haughton and Adam Phillips suggest, arrestingly, that "two hundred years after Clare's birth, Clare still speaks to us with something of the exemplary perplexity of the displaced person, of an exile within his own country."[1] While Clare usually is considered to be a poet of local relevance and distinction—the poet of Helpston, Northamptonshire, or the Midland moors and fens—he appears here not as a "village minstrel," to recall the title of his second volume of poetry, but as a "displaced person." In marshaling terms more conventionally used to describe forced migrations and survivors of atrocity, including a category—the displaced person (DP)—that emerged in the aftermath of the Holocaust, Haughton and Phillips might be perceived as offering a hyperbolic account of a minor romantic poet, one that at its worst risks trivializing a deplorable political condition in an effort at what Tom Paulin calls his redemption.[2] Yet, if we linger for a moment with this account of Clare, we can recognize—as Haughton and Phillips incisively recognize—in Clare and through his writing the existence of a position of internal exile, displacement without mobility, homelessness, and nonbelonging that remains inadequately understood, both in Clare's time and in our own.

While many scholars have focused on Clare's writing about the violence of enclosure or experience of mental illness, by framing Clare in terms of exile and displacement, Phillips and Haughton suggest that the very poet who could be taken to exemplify the convergence of romantic naturalism, emplacement, and genius, that is, a poet who wrote in a local dialect, maintained a profound awareness of the local landscape, and perceived himself as a native, is at the same time a displaced person. I propose to consider how these two conditions can converge, how a radical experience of local attachment is not at odds with displacement, and conversely, how displacement can become a form of attachment. More than this, as I will suggest, first pausing at three autobiographical texts from three decades of Clare's writing (the "Journey from Essex," the journal, and the long, apparently nostalgic poem "The Flitting"), before turning in conclusion to the ballad "Decay," Clare over and again registers an experience that occurs between versions of homelessness and a life in which belonging and attachment have a strange, nonoppositional relation to displacement. These convergences and divergences, which run throughout Clare's poetry and prose also allow us to reflect on our contemporary conditions and offer some guidance for describing and ultimately re-envisioning what, in his reading of Clare, John Barrell called "sense of place."

Phillips and Haughton refer to Clare as a displaced person, because Clare understands himself in this way. In the "Journey from Essex," Clare's account of his escape from the mental asylum, where he was an inmate, and his return to Northborough, which was not his native home, but the place to which he and his family moved ten years earlier, Clare calls himself "homeless at home." Clare returns from the asylum to a home and family he does not recognize. He expects to find Mary, whom he never married and who had died, but instead is met by Patty, his wife of many years. Yet, no evidence will persuade him of the appropriateness of his situation:

> I passed Walton & soon reached Werrington & was making for the Beehive as fast as I could when a cart met me with a man & a woman & a boy in it when nearing me the woman jumped out & fast caught hold of my hands & wished me to get into the cart but I refused & thought her either drunk or mad but when I was told it was my second wife Patty I got in & was soon at Northborough but Mary was not there neither could I get any information about her further then the old story of her being dead six years ago which might be taken from a bran new old Newspaper printed a dozen years ago but I took no notice of the blarney having seen her myself about a twelvemonth ago alive & well & as young as ever—so here I am homeless at home and half gratified to feel that I can be happy any where

For Phillips and Haughton, Clare's position, which might better be described as *internal displacement*, is intimately tied to survival, both in the immediate sense of enduring extreme poverty and perceived incarceration, and in the ontological sense of surviving one's own death, whether imagined or real.[3] Phillips and Haughton not only recognize Clare's physical displacement but also see him as one who lives and speaks beyond his time: one whose insights and observations emerge as shockingly contemporary, reflective not of a putatively romantic sense of local belonging that harbors a continuity between "nativity and nationality," but of a late modern sense of forced migrations, diasporism, and displacement, even without mobility. In their account, Clare's *physical immobility* converges with a strange *temporal mobility*, or what I have described as *survival*, and Andrew Bennett calls *posterity* one that has an uncanny resonance with Clare's own experience of Mary who he loves as if she were still alive, and as if their relationship had been borne out. Thus, when we encounter Clare in terms that are at once as old as Moses and Odysseus and as new as late modern neoliberalism, we do so because Clare, far from remaining a figure of the past and far from standing in an uncomplicated relation to his locale, is *never* at one with his own life (or, as I argued earlier, his death). When Haughton and Phillips figure Clare as bearing a voice—old, enduring, and strange—that remains uninterrupted, whether by death or by a history of reception that has often left him mute and abandoned, they also figure him as a survivor: he is out of place, but also out of time. This condition of nonidentity, at once social and existential, leads him to speak and be heard well beyond the time of his life; it is a radical localism that becomes indissociable from an experience of nonbelonging; and it even can be understood to anticipate the state of responsibility under modernity as Theodor Adorno understood it when he wrote that "today . . . it is part of morality not to be at home in one's home."[4]

When Clare's readers talk about his homelessness, they tend to evoke two apparently distinct, and even irreconcilable, experiences: the Enclosure Act of 1809 (implemented only in 1820, the year in which Clare's first volume of poetry appeared) and Clare's 1832 move from Helpston to Northborough. With the enclosure of Helpston, a new, legally imposed economic and agricultural structure emerged, a form of what in contemporary parlance we would call "development," and it transformed the actual landscape of Clare's parish. While Clare did not move at this moment, the commons was reshaped beneath him, and the effects devastated him. As John Barrell has documented, enclosure turned Helpston's open field agricultural model, built around the commons, into a closed one, newly defined by fences, boundary lines, and a reorganized distribution of activity. Barrell holds that the effects of enclosure on the lives of the poor and on Clare's poetry are more difficult to establish than Clare's most sympathetic readers have assumed. However, this imperceptibility can be understood as yet another instance of an experience of displacement that falls outside of the conventions of measurement. In other words, this is *displacement without displacement* and an experience of impoverishment that does not register on the usual scales.

The second experience of homelessness is a form of movement perceptible only when we alter our scale of measurement. Beginning in childhood, Clare took a series of small journeys, movements that from most perspectives would seem insignificant. These included a trip just across the road, which seemed to Clare like a new world, and later to London, which was in many senses a world away. But these also include actual moves—to the asylums of High Beech and Northampton, and later, in 1832, to a new home in Northborough, only a few miles from his native parish, but which, as Douglas Chambers suggests, "provided him with a metaphor of exile."[5] These apparently minor movements had major significance. Clare's intense and thoroughly detailed knowledge of his surroundings made him sensitive to changes that others might never know, and hence figured (in both the passive and active senses) an experience of loss for which there is little measure outside of Clare's work.

Taken together, we can recognize two related but nonidentical forms of displacement that occur despite Clare's complete or near immobility. Both displacement by enclosure and the experience of moving homes becomes, as these other critics recognize, an experience akin to exile, while they remain virtually invisible when considered using usual tools of perception. In the first place, Clare *did not* actually move, but the land beneath and around him was rezoned and reallocated, transforming the open space of his parish into private, more efficiently managed plots. This is a form of "displacement in place" effected by geopolitical redescription through enclosure and the formation of another version of what in the last chapter I referred to as a "zone of social abandonment." In the second place, Clare *did* move homes. His new home was less than three miles away, although, as Jonathan Bate has acknowledged, this "three miles might as well have been three hundred."[6] While the first experience involves *actually* remaining in place, the second only *appears* as remaining in place, and a closer eye, one like Clare's own eye, reveals instead the microcosmic meaning of home for one whose sense of knowledge and observation is intense, and for whom displacement can occur at an almost imperceptible scale.

When Bate registers the utter specificity of Clare's geographical attachments, which are measurable by inches and feet rather than miles, oceans, or states, he also reveals a calculus in which what, from a conventional perspective, appears as remaining in place now takes on the character of displacement or exile. This occurs when belonging relies on a meaning and level of precision imperceptible using common measures so that small moves have an apparently disproportionate effect and generate a response that seems unsupported. Displacement in this sense is a form of uncounted experience, that is, an experience that exceeds or disorients representation, causing a description of the event to appear hyperbolic or excessive, rather than properly calibrated.[7] Thus, I want to suggest that Phillips and Haughton's account of Clare as a displaced person, refugee, or exile is not simply an illustrative exaggeration (or a form of poetic license), but rather part of an effort to take seriously two experiences of displacement that fall outside of our usual schema. While Clare's experience seems slight by comparison to large-scale forced migration or contemporary experiences of internal displacement caused by famine or war, these modes of displacement, which depend upon changing attachments and hyperawareness of place, still matter, both in Clare's time and in our own.

One place where we can see the connection to our own time is in Rob Nixon's recent account of *Slow Violence and the Environmentalism of the Poor.* There, Nixon identifies "displacement in place" as a condition spawned by development, in which a "community refuses to move but, as its world is undermined, effectively becomes a community of refugees in place."[8] Engaging examples from the global South, and Jamaica in particular, Nixon sets out to describe "a more radical notion of displacement, one that, instead of referring solely to the movement of people from their places of belonging, refers rather to the loss of the land and the resources beneath them, a loss that leaves communities stranded in place stripped of the very characteristics that make it inhabitable."[9] This condition of *immobility* seems to describe Clare's experience of enclosure and its destruction of the commons: "When ploughs destroyed the green, when groves of willows fell" ("The Village Minstrel," stanza 90), "These paths are stopped—the rude philistine's thrall / Is laid upon them and destroyed them all" ("The Moors"). This is an experience that, as Phillips and Haughton suggest, left Clare a kind of refugee while remaining at home.

When Barrell sets out to better understand the effects of the enclosure of Helpston—on its population, on local conditions, and on Clare's poetry, he also discovers that the effects of the enclosure are significantly more difficult to register than Clare's earlier critics had assumed. Barrell notes the delay between enclosure's enactment in 1809 and its implementation in 1820 and goes on to acknowledge the real challenge of establishing causal relations between enclosure and violence, even when they seem intuitively justified. At the same time, he admits the difficulty of taking at face value the apparent economic value of enclosure, when the details of its effect on the lives of the poor are effaced.

Barrell thus observes that the enclosure of Helpston *might* have helped the demand for labor because, as he notes, although there was a general agricultural depression during the period that enclosure was set in place, once initiated, "the actual work of enclosing" could not be abandoned; hence, enclosure led to what in today's economy we would call "job creation."[10] But, he also recognizes, "there is quite simply not enough evidence to allow us to say for sure that, in its economic effects at least, the enclosure of Helpston was not as disastrous to the agricultural labourer as the traditional view of enclosure would suggest."[11] Here Barrell reiterates as economic history a claim that he makes earlier about poetry. Barrell hesitates before the status quo readings of Clare to suggest that there may not be enough evidence in the poetry to justify the dominant reading of enclosure attributed to it. He turns to Clare's poem *The Parish,* conventionally read "as an account of the effects of the enclosure on the social structure of Helpston" to say boldly that "in the first place *The Parish* is not much concerned with enclosure at all," and further that even what Clare says "about enclosure in the poem is ambiguous."[12] Barrell's double hesitation here—to question both the effects of enclosure and the conventions of its measurement, whether statistical or poetic—suggests not that we should *not* think about Clare and displacement, but rather that there is a long history of confusion around Clare's poetry and its politics that traditional readings of Clare, which Barrell goes far to reconsider, simply miss. One way of building upon Barrell's hesitation is to consider how Nixon's contemporary examples might help us to understand enclosure, its legacies and parallel formations, and perhaps vice versa.

In fact, the connection between the situation that Nixon patiently describes as "slow violence" (and as I have been similarly suggesting is a form of unquantifiable violence) has another, rhetorical connection to Clare's experience. When Nixon describes antithetical forms of contemporary displacement, he twice uses a phrase that could belong (and in some ways does belong) to Clare. Nixon figures the slow violence of what he calls "stationary displacement" as one in which "impoverished communities . . . have been involuntarily *moved out of their knowledge*," and he goes on to suggest that the "threat of displacement without moving . . . entails being simultaneously immobilized and *moved out of one's living knowledge* as one's place loses its life-sustaining features."[13] Staying in place is a trap, yet mobility is not an option: the movement and loss that Nixon describes coincide not with leaving one's home, but with remaining in it, much like a contemporary version of enclosure. And, like enclosure, the effects of these forms of violence are difficult to track using conventional economic or poetic strategies.

Although unacknowledged here, "getting out of knowledge," which is how Nixon describes the experience of being abandoned to a place that is undergoing a transformation that precludes the participation of its local population, could be Clare's phrase. Clare uses this sense of *knowledge* not when he speaks specifically about the enclosure, the experience in his life that most clearly resembles the experience of displacement that Nixon describes, but rather in his *Autobiographical Writings* when he recalls a childhood journey in which he sought to reach the landscape's edge, but only ended up lost in the neighboring heath, leading him to experience both the trees on the other side of the road and his own village as foreign lands. Clare describes the experience in this way: "I got *out of my knowledge* when the very wild flowers and birds seemd to forget me and I imagind they were the inhabitants of new countries" (my emphasis).[14] Clare's account of getting "out of knowledge" refers to an experience of displacement that differs significantly from the one that Nixon identifies and also from the experience of enclosure that he recounts in several poems in which he alternatively laments and contests the new signs, fences, and roads that tear up and reorient his parish, and which are typically understood as the referent when Clare writes about homelessness.[15] Nixon's work on contemporary cases demonstrates why it is so easy to confuse the two notions of homelessness in Clare's writing, for although Clare focuses his affective account of displacement in the poetry and prose devoted to minor mobility, the experience provides the terms through which the effects of enclosure can be understood, but terms that are not in fact used to describe it.

Here, Clare's experience is, at least most immediately, an effect of his intensely local awareness, as well as voracious curiosity, a mistaken understanding of geography, and an experience of disarming anxiety, rather than development or modernization. Clare arrives at this experience of losing his home on a Sunday afternoon when, consigned to the fields to tend cows and horses, he decides to follow his whim and find out whether it is possible to touch the horizon, the point at which land meets sky. Clare wanders into the heath just on the other side of the road to Peterborough, at the outer tip of Helpston. Intending to reach the end of the world, he finds himself instead in what feels to him like another country and hemisphere. He gets lost, and discovers not only that he has entered into another territory, but also that his familiar territory has become a foreign land. He explains:

> I had imagined that the worlds end was at the edge of the orison and that a days journey was able to find it so I went on with my heart full of hopes and pleasures and discoverys expecting that when I got to the brink of the world that I could look down like looking into a large pit and see into its secrets the same as I believd I coud see heaven by looking into the water so I eagerly wanderd on and rambled among the furze the whole day *till I got out of my knowledge* when the very wild flowers and birds seemd to forget me and I imagined they were the inhabitants of new countrys the very sun seemd to

> be a new one and shining in a different quarter of the sky still I felt no fear my wonder seeking happiness had no room for it I was finding new wonders every minute and was walking in a new world often wondering to my self that I had not found the end of the old one the sky still touchd the ground in the distance as usual and my childish wisdoms was puzzld in perplexitys night crept on before I had time to fancy the morning was bye when the white moth had begun to flutter beneath the bushes the black snail was out upon the grass and the frog was leaping across the rabbit tracks on his evening journeys and the little mice was nimbling about and twittering their little earpiercing song with the hedge cricket whispering the hour of waking spirits was at hand which made me hasten to seek home I knew not which way to turn but chance put me in the right track and when I got into my own fields I did not know them every thing seemd so different the church peeping over the woods coud hardly reconcile me when I got home I found my parents in the greatest distress and half the village about hunting me.[16]

When Barrell, who, like nearly all of Clare's contemporary critics, reflects upon this extraordinary passage, he suggests that Clare registers something more than just moving beyond the familiar limits of his village. Barrell first points out that "out of knowledge" is not a phrase that Clare invents. In fact Robert Bloomfield, whose work Clare admired, also uses it. However, the phrase is one to which Barrell believes Clare gives an expanded meaning.[17] He explains: "As Bloomfield uses the phrase its meaning is clear enough . . . and for Clare, too, the primary sense of the phrase is 'out of the place I knew'—the place he was familiar with and knew his way about. But it is clear too that Clare is quite conscious of asking the phrase to mean much more than that, to carry the full weight of its literal meaning—not just out of the place I knew, but out of *everything* I knew."[18] Thus, although the metaphor suggests entering into an unfamiliar place, Barrell suggests moreover that this movement is one beyond any recognition, one through which Clare ends up in a zone outside of knowledge itself.[19] Barrell goes on to point out the various ways in which Clare experiences this new world—or the world as otherwise—a transformation that renders everything different and foreign. Literally out of place, Clare is not just uprooted but also upended. He has not just traveled outside a familiar zone but has entered into a zone of confoundedness and disorientation, in which the familiar has become unrecognizable. The world has become not the realm of the knowable, but that of the unknown even when known. Indeed, while Clare never does reach the horizon in a physical sense, metaphysically speaking, he does find himself peering over the abyss, unable to return. Recalling the horizon as bearing a temporality of permanent, abyssal deferral, Clare's experience is one of the profound loss of orientation itself in a movement that can have no end because the destination is always at a distance.

Despite the fact that both Clare and Nixon describe an "out of knowledge" experience, at first glance Clare's experience is more poetic or ontological than conditioned by social or economic forces. In this sense, it seems far from the one that Nixon finds in Jamaica and elsewhere, where land is transformed beneath its inhabitants' feet, and where what Giorgio Agamben has called "the continuity between nativity and nationality" or "man and citizen," a continuity regularly associated with a conventionally romantic account of place, is broken. Yet the experience that Nixon describes in words that *could* be Clare's nevertheless resonates, as I have suggested, with the enclosure of the English countryside several hundred years earlier and the act of which Clare may be the best-known witness. Some have suggested that in reflecting upon his childhood, Clare already is accounting for an experience of enclosure as exile without return. For example, Jonathan Bate, reflecting a sentiment shared by many of Clare's contemporary readers, reads this anecdote to mean that "once a native has gone away, he can never fully return."[20] Here Bate marshals an idiom that seems better suited to cosmopolitans like Rushdie or Conrad than to the young Clare lost on the road across from his parish. Yet, when Clare's readers register the resonances between his experience as a lost child and those of geographic exile and political displacement, both his own as those experienced in other times and places, they accomplish two ends: They simultaneously conflate two distinct experiences of homelessness and register two vivid emblems of the profound, intractable interrelation between physical and psychic violence.

If the example of Clare involves two concepts of homelessness at home, one defined by enclosure and the other by small-scale, virtually imperceptible experiences of mobility that occurred throughout the course of his life and maintained the power to turn him "out of knowledge" and turn his home into a foreign land, the resonance between these two modes, at once virtually identical and sharply distinguished, is unsettling. Clare's experience of crossing the road is significant enough to lead him to feel as if he is entering "new countrys" (if not reaching the end of the world, as he had hoped), and also to cause his family and village to fear his disappearance, even his death, a fear warranted not only by the fact that there had been actual falling trees, but also by the outcome of Clare's radical experience of nonknowledge.[21] He is lost, as a child or sometimes an adult can get lost—unable to find his way home. But this is not the end of the story: Once Clare reaches home, he no longer knows that he has arrived. On the journey back to his village, Clare's home itself comes to appear as a strange, unrecognizable land. He explains that upon returning from the dark woods, his village, both on the horizon and up close, appears foreign and not his own. Even structural, human evidence—"the church peeping over the woods"—was hardly enough to prove he had returned; this visible evidence, Clare recalls, "could hardly reconcile me," and as he explains, "when I got into my own fields I did not know them every thing seemed so different."[22] Bate has suggested that when the church becomes unrecognizable, it is not because the landscape has changed, but rather, because *Clare* has changed, explaining that: "Clare spent many of his happiest days wandering alone on Emmonsales Heath. He grew intimate with the flora and fauna of Helpston and its surrounding parishes. His sense of his own identity was bounded by the horizon of his locality. To leave Helpston was to go out of his knowledge. To return was unsettling: the known and loved place seemed different. In reality, Helpston was the same. It was Clare who was different." He concludes with the formula that I cited earlier: "once a native has gone away, he can never fully return."

Here Bate moves from highly specific knowledge (that of Clare's relation to Helpston) to a general statement about the impossibility of return. Yet, what does it mean to go away? Referring neither to Clare's later visits to London nor to his entry into the imaginative universes of literary work (including that of Crusoe), the context of Bate's account of a perpetual experience of displacement is the moment at which Clare *merely* crosses the road into Emmonsales Heath. While in some empirical sense, it is true that Helpston had not changed over the course of the hours during which Clare was away, Clare, for what may be the first time, sees Helpston from another angle, both in a literal sense and in a figurative one. Clare sees it in such a way that he describes the inverse of cosmopolitanism, but one that does not send one's attachments back to the nation or to humanity rather, this experience is one of radical homelessness, even at home. More than this, in his retelling of the story Bate highlights the hyperanimation of the evening landscape, and in doing so, he evokes another sense of being out of knowledge. When he emphasizes the fluttering moth, whispering cricket, peeping church, and hour of waking spirits, Bate reminds us of another example of the strange resonance between two versions of homelessness and another name for feeling homeless at home: the *uncanny*.

In 1919, about a century after Clare's account of getting lost in the woods, Freud published his paper on the *unheimlich*, the uncanny, or—as James Strachey reminds us in a footnote, what is "literally 'unhomely.'"[23] While Freud derives his interest in the uncanny from the work of Ernst Jentsch, his aim in the essay is to provide a more complicated account of unfamiliarity, one that, drawing *first* upon linguistic evidence and *then* upon literary and anecdotal evidence, undertakes to identify the ironic or ambivalent status of a condition that cannot be sufficiently distinguished from its opposite. In Freud's account, the *unheimlich* (the strange or unfamiliar) comes to be uncomfortably like the *heimlich* (the homely, familiar, and cozy). Or, thought from the inverse position, Freud explains: "What interests us most . . . is to find that among its different shades of meaning, the word '*heimlich*' exhibits one which is identical with its opposite, '*unheimlich*.' What is *heimlich* thus comes to be *unheimlich*."[24] Clare's journey out of his knowledge has the effect *not* (or at least not at first glance) of making the strange familiar, but rather of making the familiar strange, a gesture that, as Freud's insight suggests, also makes the strange *like* the familiar. Clare's journey out of his knowledge, traces a figure-eight: a trip to the nether-lands leaves him missing his planned destination (the edge of the world), going almost nowhere (into the neighboring heath), and returning to a home in which "every thing seemed so different." Recollected many years after it actually occurred Clare's experience of unfamiliarity is out of place in a textual sense as well as a geographic one.

This division within the event also recalls the temporality of the enclosure to which, as I have been suggesting, it does not directly refer. Yet, enclosure is an experience that, like the episode that Clare remembers, bears a complex temporality, as the publication and enactment of the Enclosure Act of 1809 are separated, like this event and its telling, by nearly twenty years. While Clare's experience of the *unheimlich* has nothing to do with enclosure as such, it also simultaneously remembers and anticipates it and its divided temporality. By this I mean that the execution of the enclosure is not identical with its enactment, leading to a divided point of reference. Indeed, what is at stake here in thinking about Clare's homelessness is not only his relation to place, but resonating with Nixon's thinking of slow violence, a temporal condition that renders the place of home one's own and not one's own, unified and divided. If the enclosure is an event like those that Nixon identifies, that produces a form of displacement-in-place and "getting out of knowledge," it is also a form of violence that unsettles and redirects even those accounts to which it has no explicit relation past and future. In other words, the very act of dividing, organizing, and privatizing, an act that seeks efficiency and economy, effects not only a geographic but also a temporal disarray. Disarray here is not strictly an effect of social and economic violence, just as it may not be a strictly fundamental condition.[25]

This is true of homelessness itself when it becomes a name both for a fundamental experience of nonbelonging and for actual historical experiences of migrancy and displacement, even in place. While the urgency of political violence tempts us to dismiss the psychic or existential conditions of alienation as hyperbolic responses or merely personal, human attachments in order to rigorously describe, and make every effort to ameliorate, it, I want to suggest, as I am arguing Clare's writing suggests, that the difference between these two experiences cannot simply be established even in a body of work as apparently contestatory as Clare's own. These versions of radical homelessness remain within an untrackable yet intractable relation, a relation that has the character of the *unheimlich* itself, and they share its forms of mobile immobility and temporal disarray. My suggestion, then, is twofold. In the first place, as I already have suggested, we need to register the difference between two modes of homelessness, specifically, but not only, as they emerge within Clare's writing, calling our attention to the difference between displacement in place and minimal mobility. While the effects of the former suggest a loss that cannot be recovered because the place to which one belonged is now physically reorganized, the latter registers another form of loss, one that is the *effect* of radical attentiveness, of a belonging so intense that it turns what by all conventional accounts—national, regional—would be called one's domain into an utterly foreign land from which one then comes to experience every sector of the world as one that has generated or has the potential to generate an experience of displacement or homelessness. In this case, small differences, like a shift in perspective, can amount to toppling changes and a loss that is simultaneously the effect of recognition and misrecognition.

In the second place, there remains the relation between the aesthetics, philosophy, and experience of homelessness. While there is reason to rethink the conflation of various forms of homelessness and to distinguish between forms of loss, this rethinking can and should occur alongside a reconsideration of the relation between the performances and theories of displacement that inform them. This amounts to a redescription of the relation between the aesthetics of the uncanny, the philosophy of essential homelessness, and the theories of displacement, a reconsideration that sets out from the acknowledgement of the multiple experiences of being homeless at home and that, by way of Clare, engages in a rethinking of the assumptions of romanticism itself.[26]

Over and again, Clare's readers register his personal disorientation as an institutional one and confuse his various experiences of what he calls homelessness—getting lost as a child, moving homes, entering the asylum, escaping from the asylum, and so on—with those of economic and social displacement effected by the Enclosure Acts. To offer just one example: This is what Chambers does when he says that there is a poetic value to Clare's 1832 move for it, as I mentioned earlier, "provided him with a *metaphor* of exile," and, as Chambers continues, "a means of articulating something only glimpsed in the earlier poems and associated with lost childhood and the world before the enclosures."[27] For Chambers, the later experience of minor displacement has value as a metaphor or substitution, one that because figured as a metaphor can be taken to absorb all of the other experiences into its chain of significations. But that logic, whereby exile is merely a metaphor and whereby all of the experiences become attached through the articulation of this later event and its figuration simultaneously affirms and undoes the hierarchy of Clare's examples. Yet, is there another way to connect the experiences of psychic, aesthetic, social, and economic displacement? Surely, this transfer seems justified both by Clare's own extraordinary manner of describing a range of displacements and by the very logic of displacement and incongruity to which he bears witness when he venomously writes about the violence of enclosure. Even so, when readers take Clare's account of feeling homeless at home upon returning to Northborough after escaping the asylum to reflect the acts of geographic reinscription that he also suffered as a result of enclosure, or when they understand a personal or poetic experience of displacement as one that is *merely* continuous with forms of economic violence in the nineteenth century and the present, what they miss is the very structure of displacement that conditions the critical response. If displacement is displacement, if it reveals that home is or always can become unrecognizable, if it teaches us that the familiar is no longer distinct from the strange, the continuous logic that has been important to identifying the psychic effects of poverty, labor, and contingency is untenable. Rather, understood as reflections of the force of a profound and disorienting discontinuity, one that unsettles experience itself, Clare's accounts of homelessness register, in their very eloquence, the failure to remember personal or social histories. These histories come to be mere reflections of one another, or, put another way, they reflect an experience of homelessness in which the uncanny becomes both a profound way of registering historical power and a failed means of arresting an event that belongs to a particular time and a particular place.

The anecdote from the early 1820s has been in the background of my thinking here. In it, when Clare recalls a moment from his childhood, he also simultaneously remembers and anticipates, which is to say, *repeats* even in advance of their taking place his other experiences of displacement, including those tied not to childhood curiosity, but to enclosure. The challenge of repetition, however, is that the temporalities of enclosure and of this recollection render any decision about the generative event, that is, about whether enclosure is the source of Clare's experience of homelessness, or whether a more fundamental or existential experience of homelessness (and eventually madness) is at the root of his suffering, an impossible task. While enclosure uproots Clare, renders him homeless at home, and achieves the condition that Nixon describes as signified by getting out of one's knowledge, it is also a childhood accident that leads Clare to provide the account through which the enclosure's particular brand of displacement as "getting out of one's knowledge" while remaining in place can be understood.

Likewise, it is a later experience of moving to a new home—and then returning to this place, that is neither one's native home (Helpston) nor where one lives (the asylum)—as a home that provides the language through which the experience of enclosure has been understood. In other words, despite prevailing accounts of Clare, and despite Clare's evocation of enclosure, homelessness becomes an example of a catachrestic or improperly counted experience. It is an experience of poetry, and it shows how poetry becomes an experience of homelessness. This does not mean that Clare's poverty and his social condition are irrelevant. Neither does it mean that the enclosure of Helpston or the experience of the asylum as a shelter that is not a home, organized precisely to solve the challenge (real or imagined) of not being able to return to one's home do not matter here. It does not simply stage a flattening equivalence between all of these intricate experiences. Rather, each of these experiences simultaneously reveals and generates a series of displacements—registered in experience, in poetry, and in criticism.[28]

In order to capture some sense of this experience—one that is and is not identical to a social condition, one whose geography is and is not an effect of enclosure, I want to turn from Clare's autobiographical fragment to the archive of his poetry, and in particular to focus on two poems that appear in the 1835 volume *The Rural Muse*, the last volume that Clare published in his lifetime.

Clare wrote the poem initially called "On first leaving the cottage of my youth" to commemorate the painful experience of moving from his native home. The poem later took on the title "The Flitting," which is what such a move is called.[29] A *flitting* also refers to a range of experiences of mobility, whether in a metaphorical sense of changing one's position of mind or a literal one of scurrying, whether to move temporarily, like a bird or insect flying away, or more permanently as a final departure, whether in disappearance or death. The poem begins with a lament for what has been lost in the move from Helpston to Northborough, largely figured in terms of color: "green fields," the hazel's happy green," and "The bluebell's quiet-hanging blooms." Yet, this physical displacement is registered as a temporal one. The flitting leaves Clare in a new landscape, one in which it is not *he* who is a stranger, but rather *summer* itself has become unrecognizable to him. The poem begins:

> I've left my own old home of homes
> Green fields & every pleasant place
> The summer like a stranger comes
> I pause & hardly know her face.
> I miss the hazels happy green
> The bluebells quiet hanging blooms
> Where envys sneer was never seen
> Where staring malice never comes[30]

For Clare, the land bears a face, it is the face—or in this case, the defacement—of time as marked by the seasons. In the opening stanza, Clare refers to his cottage not merely as a home, but hyperbolically, as a "home of homes," the genitive here suggests not belonging, but exemplarity.[31] While "of" is the preposition used to state attachment, genealogy, and origin (e.g., as Clare writes later in the poem "tenants *of* an ancient place" and the poem's initial title refers to "the cottage *of* my birth"), here it is used to register incomparability, that which stands alone and refers only to itself.

This is the crisis that the poem initially reflects—the home that is most one's own also no longer is where one lives. The effect is so severe that even the various scenes that would provide comfort or recognition leave Clare instead displaced and demeaned. The operative trope in this opening stanza is personification, specifically a prosopopoeia or facing, even as it is used to account for an experience of depersonification or displacement that is registered in the act of pausing that defines Clare's relation to this new scene. It is ultimately not that the place leaves him permanently thwarted, but rather it operates like an interruption that dashes his knowledge and prompts—like the experience of the uncanny or the address "To Mary"—a delayed recognition. Once the faces return—faces that are attributed to the fields and flowers—they are vicious, but only implicitly so. They—"envy's sneer" and "staring malice"—were *never* a part of Clare's "home of homes," although by implication, rather than assertion, they are what he experiences here; they are unfamiliar, but they are also a response to the unfamiliar.

The conjunction of loss, displacement, and near unrecognizability appears in another poem included in *The Rural Muse*, "Decay: A Ballad." While the poem is formally very different from "The Flitting," relying upon the internal repetitions of the folksong to account for an experience of loss, like "The Flitting," it also tells a story of homelessness and displacement. The object of decay in the poem is *poesy* itself, and this is remarked through a set of variations on the ballad's opening line, "O poesy is on the wane."[32] The poem's first stanza reads:

O poesy is on the wane,
For fancy's visions all unfitting;
I hardly know her face again,
Nature herself seems on the flitting.
The fields grow old and common things—
The grass, the sky, the winds a-blowing
And spots where still a beauty clings—
Are sighing "Going! all a-going!"
O poesy is on the wane,
I hardly know her face again.[33]

Here, as in "The Flitting," Clare registers an experience of nonrecognition. Both poems also open with a loss tied to displacement: In one, he moves house; in the other, poesy is in decline. Yet, the loss that "Decay" archives, even as it describes a transformed terrain, is not a physical change, but an aesthetic one. What is lost is a way of seeing—or of reading. As Clare continues:

The sun those mornings used to find
When clouds were other-country mountains
And heaven looked upon the mind
With groves and rocks and mottled fountains.
These heavens are gone, the mountains grey
Turned mist, the sun a homeless ranger
Pursuing on a naked way
Unnoticed like a very stranger.
O poesy is on the wane,
Nor love nor joy is mine again.[34]

Here, the mountains turned to mist, were not initially mountains, rather they were clouds, read as mountains, and it was the sky that was seen, poetically, as belonging to the earth. If the mountains have become mist, if *these* heavens are gone, it is because they were never there, because they were "fancys visions." While the poem *seems* to track the decay of a particular place ("the brambles dwindled to a bramble"), in fact it tracks the decay of a mode of perception, the failure to continue to see a common stream as more gorgeous and more significant than a manicured landscape; the inability to see the heavens as a kingdom rather than a mist; the sun as a sovereign, rather than a "homeless ranger."[35] In other words, it accounts for the loss of a delusion. Again, what is curious and disorienting about the poem is that it is an account *not* of a physical place but of perception; it is an account of decay figured as a changing *terroir*, but in which that change turns out to be one of lost ideologies and thwarted systems of perception ("mushrooms they were fairy bowers, /Their marble pillars overswelling").[36] If this is a rude awakening, awareness, or "undeception" figured as loss, decline, and decay, and the longing for a past naïveté to which one no longer has access, the narrative of loss is borne by repetition and internal difference.

Throughout "Decay," the rhyme on *wane* is *again*, whose meaning shifts during the course of the ballad from implying repetition (this is happening *again*) to impossibility (this will never happen *again*) to desire (I wish it would happen *again*). In other words, the ballad comes to show that *again* for all its repetition might not mean *again* at all, and that the only thing about which we can be certain is the repetition itself rather than how to read it. But this kind of repetition—the recurrence of a term, the recurrence of the term of recurrence, but with a difference—is not the same as decay, especially not the loss of an illusion figured as decay. If the earlier passage from Clare's *Autobiographical Writing*, reveals an instance of minor displacement and "getting out of one's knowledge" that renders the familiar strange, in "Decay" and "The Flitting," these moments of not fully recognizing the familiar—*summer*, *poesy*, even the stable meaning of the term *again* suggest a different kind of loss—and recovery. In fact, "Decay" may be read as an account of the recovery of the common as common, as a merely, unpoetic place. Indeed, it is the "common things" that issue this lament, not for themselves, but for a lost vision. If in "To Mary," "common air" suggests a form of uncanniness, the immateriality of thought and the lingering tales that inhabit the wind, here, when the common things whisper, it is not to mark an excessive presence, the life of what does not exist but cannot be lost, but rather to announce poesy's apparent disappearance. Put another way, while "Nature herself seems on the flitting," she only *seems* so. The poem instead reveals the various ways that a prosaic scene endures, the way that decay—the disappearance of fantasy (or fancy), belief and deception—might lead to a sadder and sorrier existence, but that the flowers, streams, and even the sun still remain, again and again and again.

When Barrell reads "The Flitting" as an account of awakening, he understands the poem to register and ultimately commemorate the possibility of learning to love the natural world in its generality. The poem archives Clare's hard-won ability to turn the limited feeling of attachment to a particular place that he perceived as home into a feeling of universal love for nature. For Barrell, the poem marks the birth of a cosmopolitan Clare. Partway through the poem, Clare's heart begins to turn as he discovers that what he thought was a feeling of purely local attachment transcends local experience. This turn happens when he recognizes the "simple weeds" that grow in Northborough. He recognizes them as familiar to Helpston, realizes that he can love them, that his love for the particular is for every single particular, not just one, and that this is not a new position but a better way of understanding how he always has felt, and how he recurrently feels in the spring. What seems to be a moment of discovery turns out to be an instance of revelatory repetition. He writes:

E'en here my simple feelings nurse
A love for every simple weed
& e'en this little shepherds purse
Grieves me to cut it up—Indeed
I feel at times a love & joy
For every weed & every thing
A feeling kindred from a boy
A feeling brought with every spring[37]

So the transition begins with a weed, that is, a plant that grows out of place, and more than this, that belongs to no place, neither to Northborough nor to Helpston. The realization turns the first experience of summer's unrecognizable face into a spring that is exactly the same as it ever has been and yet radically transformed. For all of the universality that this experience seems to suggest, it also turns out that it too depends, almost too predictably, on a familiar cosmopolitan logic that requires identity and identification.[38] The spot may be strange, but its inhabitants are the same everywhere, and this discovery leads to a new form of strangeness in which Clare has a new feeling of attachment made possible once he discovers that these weeds are the same as those with which he lived. In other words, that they are his neighbors. Clare continues:

& why—this shepherd's purse—that grows
In this strange spot in days gone bye
Grew in the little garden rows
Of that old hut now left—& I
Feel what I never felt before
This weed an ancient neighbour here
& though I own the spot no more
Its every trifle makes it dear

The ivy at the parlour end
The woodbine at the garden gate
Are all & each affections friend

That renders parting desolate
But times will change & friends must part
& nature still can make amends
Their memory lingers round the heart
Like life whose essence is its friends

Time looks on pomp with careless moods
Or killing apathys disdain
So where old marble citys stood
Poor persecuted weeds remain
She feels a love for little things
That very few can feel beside
& still the grass eternal springs
Where castles stood & grandeur died.[39]

This is how the poem ends, and it seems to issue two claims. First, the weeds remain, even when all else has gone, even when the great cities, even civilizations have fallen and second, these "poor persecuted weeds," these lone survivors nurture time's love, and more than this, their temporal displacement (or survival) is correlated with a particular experience of belonging: they are precisely what is of the earth that is unloved, unseen, and, most important for the purpose of this thinking, what is experienced as nonbelonging or as being proper to no place. It is the appearance of a familiar weed, shepherd's purse, in this new place that allows Clare to reconnect with his love of the land. This is the same love he felt earlier, insofar as it is tied to the minor, and to use his own words, the "Mild and bland." Yet, it is also a new form, insofar as its source is now experienced not as what marks a particular place, but as what is out of place both at home and elsewhere. Once again, as in childhood, Clare finds himself out of knowledge and experiencing both the neighboring land and his home as a foreign place, and similar in their shared strangeness.

But what does it mean to imagine a world in which the weeds are all that remain? A world in which sublime stories and built environments are merely fleeting by comparison to the weeds? These weeds expose an intensely local sensibility as an ironically diasporic one, for what binds us to a place and to belonging is the ubiquity of that which remains uncontained and whose attachment is as threatening as it is resilient. Moreover, these weeds, for Clare, do not replace homelessness with belonging, nor are they simply "nonnatives," but they replace personal and historical experience with a transhistorical, common one.

These lingering weeds lead me to conclude with another contemporary scene, neither from Clare's England, nor from Nixon's Jamaica, but from Detroit, the city in which I, like my parents and maternal grandparents, was born (and from which they fled to end up in a new suburban landscape that I both do and do not call home). In fact, these reflections on homelessness are themselves a displacement, a thinking about the forms of alienation and belonging that are honed in suburbs and other places where belonging and non-belonging are not simply opposed; places in which the forms of identity and attachment that rely upon diasporism are cultivated, places in which no one quite belongs and in which there are plenty of houses, but not many weeds.

In 2009, James Griffioen, a lawyer turned photographer and post-college Californian turned Detroiter, began to document a phenomenon—the effect of radical poverty, displacement, and homelessness—that he called "feral houses." He writes:

> Our word feral comes from the Latin root *fera*, or "wild beast," but it also has a connection to another Latin word, *feralis*, literally: *belonging to the dead*.
>
> I've seen "feral" used to describe dogs, cats, even goats. But I have wondered if it couldn't also be used to describe certain houses in Detroit. Abandoned houses are really no big deal here. Some estimate that there are as many as 10,000 abandoned structures at any given time, and that seems conservative. But for a few beautiful months during the summer, some of these houses become "feral" in every sense: they disappear behind ivy or the untended shrubs and trees planted generations ago to decorate their yards. The wood that framed the rooms gets crushed by trees rooted still in the earth. The burnt lime, sand, gravel, and plaster slowly erode into dust, encouraged by ivy spreading tentacles in its endless search for more sunlight.
>
> Like some of the dogs I've seen using these houses as shelter . . . these houses are *reverting to a wild state, as from domestication*, a word derived itself from *domesticus* (the Latin for *belonging to the domus, or house*). Now these houses are *feralis*. They belong only to the dead.[40]

In a later post, Griffioen continues:

> A feral house exposes its vulnerable material core: concrete, wood, brick, mortar. Every home, even yours, will one day be broken down to such brute matter. In nature, you may not pay much attention to a stand of pine, or native limestone, clay, shale, or stone. But in a feral house you see these materials lain bare as they are brought back to nature. We see that these raw elements of nature underlie our most potent symbols of civilization.[41]

In some sense, these houses teach us to see and feel like Clare. They embody the understanding at which he arrives when he concludes "The

Flitting" with a sentiment that is no longer local and temporal, but general and eternal, yet whose eternity, is that of weeds:

So where old marble citys stood
Poor persecuted weeds remain
She feels a love for little things
That very few can feel beside
& still the grass eternal springs
Where castles stood & grandeur died[42]

Griffioen suggests that these lost houses allow us to see the natural world anew, and in this sense they give another meaning to Clare's phrase "homeless at home." His camera takes us to the common, allows us to see the way in which a home can become strange, and to imagine, while presenting only the weeds, the profound despair, loss, and decay that these places hold. If Griffioen points us to the beauty of these spaces, he also presents us with a scene of destruction and abandonment that cannot fully be perceived.

In "The Flitting," Clare suggests that leaving—and losing—his home also allowed him to see the world anew, to recognize ivy, woodbine, and grass, not in their locality and specificity, but in their creeping excess, ubiquity, and survival, in their uncomfortable combination of resilient attachment and nonbelonging. This is not simply the substitution of one sort of love and attachment for another, but rather, one experience of nonbelonging and homelessness for another. And this means that what is at stake in thinking about homelessness is not reducible to enclosure or madness or nativism, to debt, the rise and decline of American industrialism, or pervasive inequality. It also invites us to take this feral condition, which Clare shows us is tied up with poetry, as the source of another way of thinking about romanticism and its legacies. This is a version of attachment that does not follow the logic of naturalization assumed in "A poet is born not made," but rather draws from the recurring and inevitable displacements Clare's own vision of his survival promises.

When Hannah Arendt undertakes to describe the radical shift in the political conception of rights from inheritance to an eighteenth-century notion of ahistorical, natural, inalienable rights to a twentieth-century conception of rights "guaranteed by humanity itself," she holds that this shift also coincides with a new possibility of homelessness and statelessness experienced first in the aftermath of the First World War, and later in the Holocaust and refugee crises that crossed Europe and the Middle East in its aftermath. Yet, Clare's experience of homelessness, fostered by an excessive attachment to place that reveals a scene of minor, unremarkable movements as the occasion of profound displacement, tied to a parish, if not a state, has an outcome that initially seems to resonate with the one that Giorgio Agamben will understand Arendt to open. For Agamben, attention to the radical transformation in the conception of rights, nation, and nature, which, as we saw earlier, he associates with romanticism and postromanticism, leads to a reconception of the place of the refugee as the source of renewed thinking of politics. Agamben finds in Arendt's account of statelessness and human rights, whereby "the conception of human rights, based upon the assumed existence of a human being as such, broke down at the very moment when those who professed to believe in it were for the first time confronted with people who had indeed lost all other qualities and specific relationships—except that they were still human" the emergence of what he calls "a limit concept that radically calls into question the fundamental categories of the nation-state, from the birth-nation to the man-citizen link."[43] Agamben goes on to find in the refugee as the possibility of the emergence of a new political order, one that simultaneously exposes and disturbs the conjunction of life and law:

> The refugee is the sole category in which it is possible today to perceive the forms and limits of a political community to come. Indeed, it may be that if we want to be equal to the absolutely novel tasks that face us, we will have to abandon without misgivings the basic concepts in which we have represented political subjects up to now (man and citizen with their rights, but also the sovereign people, the worker, etc.) and to reconstruct our political philosophy beginning with this unique figure.[44]

What I have been suggesting is that there is another figure who calls our attention, one who might be incorporated by the logic of the refugee who Agamben, following Arendt, understands to expose the destiny of biopolitics. But this other figure cannot simply be absorbed by biopolitics as we have known it. While romanticism, and the example of Clare in particular, would seem to deliver an exemplary version of the conjunction of man and poet, nature and poetry operating in perfect parallel with the conjunction of man and citizen, nature and nation, each time that we discover this convergence—in all of the examples of nativity that we read through his grave, his language, his identity, and finally his attachment to place—it turns out to reveal a position that poetry already has divided.

This originary, internal division, this recovery of the various forms of nonopposition between identity and nonidentity, belonging and displacement, presence and absence, understanding and delusion that run throughout an oeuvre that would seem to exemplify a certain notion of language and life or poetry and nature, is what I have called biopoetics. This biopoetics opens up another narrative of place and displacement, home and homelessness, one that continues to be legible in the accounts of immobility that Nixon describes in *Slow Violence* and in the feral houses Griffioen finds in Detroit. It demands a history of minor experiences and a mode of reading attentive to persistent displacements and varieties of nonbelonging. It is both a condition and a method that traces the constellation I have been drawing from Clare's various accounts of homelessness to the emergence, as we speak, of a postindustrial city that Clare's poetry of survival in displacement continues to teach us to read.

Coda: The Reading of Life

I end this story with Detroit, because this is where it began. It was there that I began to think about how John Clare's poetry might help us to read about the quotidian forms of displacement in U.S. cities and suburbs, just as it might help us to think about radical homelessness or the emergence of biopoetics. In other words, I end it here in order to demonstrate what is at stake in this book. If these contemporary scenes seem to bear almost no resemblance to romantic disagreements about poetic genius or to the culmination of arguments about how to edit the work of an uneducated, yet well-read laboring class poet, I imagine them flickering and reflecting one another in a constellation like those Walter Benjamin described as the vehicles of a new form of minor history. This is a history of the forgotten, the forgetting of Clare but also the forgetting—one that we are living with right now—of reading itself.

I opened by calling for a return to close reading. I conclude by suggesting that close reading (or *just* reading) might still lead us to think anew about contemporary scenes of violence and loss, identity and belonging, survival and viability. When Benjamin saw the constellation—or allegory—as a form of redemption, it was not in the sense of a conventional dialectic or overcoming, but rather as a way of holding two distant and complex moments in mind, of letting them hold one another and save each other. Hiding within this book's subtitle stand the names of the other readers who have accompanied me (or whom I have accompanied). For me the constellation—the name—that Clare holds and even saves for us is that of Paul de Man. While to talk about "redeeming" Paul de Man sounds like the work of apology and denial, a matter that I have taken up elsewhere, the kind of Benjaminian redemption that I am interested in here is what in the opening of Chapter 1 I described as "letting a little life" into de Man's work. If, as I read him, de Man was relatively *uninterested* in life, I want to suggest that this indifference, repeated over and again by his admirers and detractors, is also what makes his work seem so desperately lifeless, so robotically linguistic, so untimely, and I am arguing, despite all of this so filled with possibility. It may offer the key or lever that allows us to live with and in a world where *questions* of viability and nonbelonging—questions of language, politics, poetry—may be all that endure. In this world, untimely reading remains a condition of possibility of thinking today.

Sometimes when I think about the nature of this thinking, I am inclined to call it "posttheory." To some, this might sound like a kind of overcoming that leads us into the domain of history or description, statistics or appreciation. Instead, I think of it in terms of the absorption of theory and its ways of reading. I think of it as a way of engaging with literature and culture that cannot do without theory, but that presents another relation to theory than the one that predominated in the last century and early years of this one. This is what I have undertaken to perform. In some sense, this reading is the strongest affirmation of the impact of what still goes by the name of theory that I can imagine, but it is unfamiliar and strange, lateral and sometimes subversive. In other words, it is a bit like John Clare himself.

Notes

Introduction: The Life of Reading

1. Michel Foucault, *"Society Must Be Defended": Lectures at the Collège de France, 1975–1976* (New York: Picador, 2003), 239, 240.

2. Ibid., 262.

3. Denise Gigante, *Life: Organic Form and Romanticism* (New Haven: Yale University Press, 2009), 3.

4. For further discussion of *Frankenstein* and biopolitics, see the chapter "Testimony and Trope in Frankenstein" in my *Romanticism After Auschwitz* (Stanford: Stanford University Press, 2007). While most examples of lyric life are not concerned with species, *Frankenstein* is a notable exception, since it is precisely man as species and its future that are at issue throughout the novel. For a discussion of the difference between a power of life and a power over life, see Roberto Esposito, *Bios: Biopolitics and Philosophy* (Minneapolis: University of Minnesota Press, 2008).

5. See Lauren Berlant, "Slow Death (Sovereignty, Obesity, Lateral Agency," *Critical Inquiry* 33 (Summer 2007): 754–80; Eric Santner, *The Royal Remains: The People's Two Bodies and the Endgames of Sovereignty* (Chicago: University of Chicago Press, 2011).

6. See Cynthia Chase, *Romanticism* (London: Longman, 1993); Marc Redfield, *The Politics of Aesthetics: Nationalism, Gender, Romanticism* (Stanford: Stanford University Press, 2003). See also the work of Walter Benjamin, which offers the most persuasive and sweeping account of this historical model and its allegorical structure. See "Theses on the Philosophy of History" in *Illuminations* (New York: Schocken, 1969).

7. The play between instance and instant here is indebted to Maurice Blanchot, *The Instant of My Death*. "Contemporary Romanticism" is a phrase that I borrow in part from Forest Pyle and Jacques Khalip.

8. See the lecture of March 17, 1976, where Foucault apologizes for "this long digression on biopower." *Society Must Be Defended*, 254.

9. "Uncontained" is Carol Jacobs's term; see her *Uncontainable Romanticism: Shelley, Brontë, Kleist* (Baltimore: Johns Hopkins University Press, 1989). This point can be read to resonate with one of Agamben's criticisms of Foucault, that what he identifies as a modern emergence in fact has an ancient origin. See *Homo Sacer: Sovereign Power and Bare Life* (Stanford: Stanford University Press, 1998).

10. Foucault is of course aware of this uncontainment, and his example of Franco's death "and the symbolic values it brings into play" in the lectures of 1975–76 offers one example. *Society Must be Defended*, 248.

11. See, for example, Frederick Burwick's *Poetic Madness and the Romantic Imagination* (State College: Pennsylvania State University Press, 1996).

12. Giorgio Agamben, *Means Without End: Notes on Politics*, trans. Vincenzo Binetti and Cesare Casarino (Minneapolis: University of Minnesota Press, 2000), 66.

13. See Franco Moretti, *Graphs, Maps, Trees: Abstract Models for a Literary Theory* (London: Verso, 2005) and *Distant Reading* (London: Verso, 2013); and the 2009 collection of essays published in *Representations* under the title "Surface Reading."

14. Although I continue to insist upon this paucity, I would be remiss were I not to mention the important recent work of Paul Chirico, Mira Gorji, and Ross Wilson, the latter of whom has devoted an exceptional—in all senses—essay to "Clare's Indistinct Array" in *Romanticism* 17 (2011): 148–59. Sarah Zimmerman also focuses on Clare in a chapter of *Romanticism, Lyricism, and History* (Albany: SUNY Press, 1999), and David Collings's unpublished essay on Clare and Nancy is an excellent counterexample to the trends that I identify.

1. The Viability of Poetry

1. See Joanna Ball's discussion of the asylum and its commitment to a "gentle" system of care in "'The Tear Drops on the Book I Read': John Clare's Reading in the Northampton General Lunatic Asylum, 1841–1864," *Wordsworth Circle* 34, no. 3 (Summer 2003): 155–58. See also Jonathan Bate's description of the Asylum's "liberal regime" in *John Clare: A Biography*, 469.

2. For a psychoanalytic and rare theoretical account of Clare's love and poetry, see Sigi Jöttkandt, *First Love: A Phenomenology of the One* (Melbourne: Re-Press, 2010). Jöttkandt recalls that Clare believed he was confined in the asylum because of his polygamy.

3. See the "Journey out of Essex" included in *John Clare by Himself*, ed. Eric Robinson and David Powell (Manchester: Carcanet, 1996), 265.

4. Ibid., 264.

5. See the "Asylum Observations" from Northampton, where he writes: "God almighty bless Mary Joyce Clare and her family now and forever—Amen; God almighty bless Martha Turner Clare and her family now and forever—Amen." In ibid., 266.

6. Jonathan Bates, *"I Am": The Selected Poetry of John Clare* (New York: Farrar. Straus & Giroux, 2003), 273.

7. Paul de Man, *The Rhetoric of Romanticism* (New York: Columbia University Press, 1984), 74.

8. Barbara Johnson, *A World of Difference* (Baltimore: Johns Hopkins University Press, 1987), 184.

9. For more recent examples that engage de Man to consider questions of capital, see Jennifer Bajorek, *Counterfeit Capital: Poetic Labor and Revolutionary Irony* (Stanford: Stanford University Press, 2009), and Anna Kornbluh, *Realizing Capital: Financial and Psychic Economies in Victorian Form* (New York: Fordham University Press, 2014).

10. Johnson, *A World of Difference*, 184.

11. "Hinge" is of course etymologically related to "hang," which is the topic of de Man's reading of Wordsworth in "Time and History in Wordsworth" and Wordsworth's reading of Milton in his 1815 Preface, and which is one of the occasions that leads de Man to think about Wordsworth not as a poet of nature, but of temporality, or what I would call, combining questions of life and time, survival. See *Romanticism and Contemporary Criticism: The Gauss Seminar and Other Papers* (Baltimore: Johns Hopkins University Press, 1993), 8ff.

12. Again, this hanging and hinging for Wordsworth and de Man describes figure as simultaneously a condition of possibility and impossibility or connection and suspension.

13. As Catherine Mills has shown, the use of 4D sonograms, that is the rhetoric of images, in antiabortion materials has as its aim the production of the fetus as a face-bearing entity—as a person or personage. See *Futures of Reproduction: Bioethics and Biopolitics* (Dordrecht: Springer, 2011).

14. Johnson, *A World of Difference*, 189.

15. Agamben suggests that Foucault does not do this sufficiently. Penelope Deutscher also persuasively considers the reasons why Agamben never takes up the matter of abortion, and these reasons, in part the complexity of analogizing abortion and the Holocaust, an analogy that is the bread and butter of the anti-abortion movement is something one would not want to touch. See "The Inversion of Exceptionality: Foucault, Agamben, and 'Reproductive Rights,'" *South Atlantic Quarterly* 107, no. 1 (2008): 55-70.

16. Johnson, *A World of Difference*, 190.

17. Ibid., 191.

18. This is a prescient precursor to Agamben's account of the indistinction between the two figures of exception: the sovereign and so-called bare life. See Agamben, *Homo Sacer: Sovereign Power and Bare Life*, trans. Daniel Heller-Roazen (Stanford: Stanford University Press, 1998). It is worth noting that Johnson does not distinguish between embryo and fetus, a distinction that is typically linked to gestational age. In this analysis, the distinction is not especially meaningful, since what remains at stake is viability or the indistinction between living and nonliving, which is at issue whether the being is more or less than eight weeks in utero. See also Agamben on infants and infancy: *Infancy and History: On the Destruction of Experience*, trans. Liz Heron (London: Verso, 2007).

19. Deutscher, "The Inversion of Exceptionality," 58. For a cultural and political account of this condition, see Lauren Berlant's observation about the post-Reaganite United States: "a nation made for adult citizens has been replaced by one imagined for fetuses and children," in *The Queen of America Goes to Washington City* (Durham, N.C.: Duke University Press, 1997), 1.

20. Johnson, *A World of Difference*, 198.

21. For another account of the centrality of the mother to romantic subjectivity, see Cathy Caruth, "Past Recognition: Narrative Origins in Wordsworth and Freud," in *Empirical Truths and Critical Fictions: Locke, Wordsworth, Kant, Freud* (Baltimore: Johns Hopkins University Press, 1991).

22. These formulations evoke Agamben's account of *Homo sacer* as a life that can be killed but not sacrificed. Again, see Deutscher for a speculative discussion of Agamben's avoidance of abortion.

23. Of course, Wordsworth had a strong sense of the lyric subject as child.

24. Jöttkandt, *First Love*, 119.

25. See for example "The Moors": "Where swopt the plover in its pleasure free,/ Are vanished now with commons wild and gay / As poets' visions of life's early day."

26. And on the intake form at the asylum, Clare's condition is understood to be associated with his "years addicted to poetical prosing." Quoted in Bate, *John Clare*, 466.

2. The Origins and Ends of Poetic Genius

1. Cited in Mark Storey, *Clare: The Critical Heritage* (London and Boston: Routledge & Kegan Paul, 1973), 168.

2. Evelyn Fox Keller has shown that the understanding of nature and nurture as opposed is a much later understanding than historians, as well as psychologist Steven Pinker, have assumed. While she admits that the two terms are treated in relation to one another, it is not until Galton's *Hereditary Genius* that the two are figured in opposition. See chapter 1 of *The Mirage of a Space between Nature and Nurture* (Durham, N.C.: Duke University Press, 2010).

3. Storey, *Clare*, 169.

4. Ibid.

5. John William Tibble and Anne Northgrave Tibble, *John Clare: A Life* (Totowa, N.J.: Rowman and Littlefield), 31.

6. See Grainger's description of this: *The Natural History Prose Writings of John Clare*, ed. Margaret Grainger (Oxford: Clarendon Press, 1983), 165.

7. Ibid., 181.

8. Ibid., 188.

9. William Ringler, "Poeta Nascitur Non Fit: Some Notes on the History of an Aphorism," *Journal of the History of Ideas* 2, no. 4 (October 1941): 501.

10. Coleridge, quoted in ibid., 503.

11. Can this "natural inspiration" be understood as plasticity? For recent discussions of plasticity and philosophy see Catherine Malabou, *Plasticity at the Dusk of Writing: Dialectic, Destruction, Deconstruction*, trans. Carlyn Shread and Clayton Crockett (New York: Columbia University Press, 2009).

12. Frederick Martin, *Life of John Clare* (London: F. Cass, 1964), 5–6.

13. Quoted in Hugh Haughton, Geoffrey Summerfield, and Adam Phillips, *John Clare in Context* (Cambridge: Cambridge University Press, 1994), 262; see also Mark Storey, *Clare: The Critical Heritage* (London and Boston: Routledge & Kegan Paul, 1973, 248.

14. Eric Robinson, John Clare, and David Powell, *John Clare by Himself* (Ashington, England: Mid Northumberland Arts Group, 1996), 183.

15. Jonathan Bate, *John Clare: A Biography* (New York: Farrar, Straus & Giroux, 2003), 269.

3. Can the Poet Speak?

1. John Ashbery, *Other Traditions* (Cambridge, Mass.: Harvard University Press, 2000), 5.

2. Ibid., 19.

3. Ibid., 2.

4. See Geoffrey Grigson, ed., *Poems of John Clare's Madness* (London: Routledge, 1949). See also Fredrick Burwick, *Poetic Madness and the Romantic Imagination* (University Park: Pennsylvania State University Press, 1996).

5. Tim Chilcott, *A Publisher and His Circle: The Life and Work of John Taylor, Keats's Publisher* (London: Routledge, 1972), 96.

6. *Poems by John Clare*, ed. Arthur Symons (London: Henry Frowde, 1908), 4. See also Paul Chirico, "Authority and Community: John Clare and John Taylor" in *Authorship, Commerce and the Public: Scenes of Writing, 1750–1850*, ed. E. J. Clery, Caroline Franklin, and Peter Garside (London: Palgrave Macmillan, 2002), 84–99.

7. Eric Robinson and Geoffrey Summerfield, "John Taylor's Editing of Clare's *The Shepherd's Calendar*," *The Review of English Studies* 14, no. 56 (November 1963): 359–69.

8. Ibid., 359.

9. Jonathan Bate, ed., "*I Am*": *The Selected Poetry of John Clare* (New York: Farrar, Straus & Giroux, 2003), xix.

10. John Clare, *Poems of the Middle Period, 1822–1837*, ed. Eric Robinson, David Powell, and P.M.S. Dawson (Oxford: Clarendon Press, 1996), 1:xiii.

11. Simon Kövesi also introduces "translation" as a term in this context when he writes: "The Oxford editors' claimed intention is to transcribe Clare's manuscripts exactly as the poet wrote them. But this cannot be true. In editorial terms, the action of 'transcription,' carries with it a transformative, a transitional, and a translatory effect. Transcription involves activity, not passivity. . . . 'Transcription' therefore demands value judgements in each choice of textual variant, or more politely put, requires the enactment of editorial 'discernment.'" Simon Kövesi, "Beyond the Language Wars: Towards a Green Edition of John Clare," *John Clare Society Journal* 26 (2007): 66–67. Although Kövesi seems to consider questions of ethics (judgment, value) here, my own discussion of the ethical claims that ground the various editorial projects follows a somewhat different trajectory.

12. See, for example, Eric Robinson's more recent discussion of editing Clare's misspellings: "There are a few misspellings of the sort we sometimes see in our students' essays such as 'where' for 'were' and vice-versa, or 'their' for 'there.' They should not cause much difficulty in understanding Clare." Eric Robinson, "Editing Clare: Words" *Wordsworth Circle* 34, no. 3 (Summer 2003): 141.

13. Bate, *"I Am,"* xix.

14. Jonathan Bate, *John Clare: A Biography* (New York: Farrar, Straus, and Giroux, 2003), 564.

15. J. L. Cherry, *Life and Remains of John Clare, "The Northamptonshire Poet"* (Northampton: J. Taylor & Son, 1873), vii.

16. Arthur Symons, *Poems by John Clare* (London: H. Frowde, 1908), 23. Here Symons refers to the manuscripts to which Cherry had access.

17. Robinson writes: "The state of Clare's manuscripts is to us still a cause of wonder. There are difficulties (common to the manuscripts as a whole wherever found, in whatever period of Clare's creativity) such as Clare's cramped handwriting with its special quirks, making it impossible always to distinguish between words ending in 'y' and 'ing' or those beginning with 'sn' or 'm,' such as 'snow' and 'mow'; the endless erasures created not only by crossing out words and letters or by writing through them, but also by chemically fading them (the Peterborough manuscripts used to smell strongly of vinegar like fish-and-chip newspaper) when in ink, or by erasures with bread, penknife, or abrasives when written in pencil." Clare, *Poems of the Middle Period*, xiii–xiv.

18. Kövesi, "Beyond the Language Wars," 64.

19. On aesthetic ideology, Andrzej Warminski reminds us that for de Man, "rhetoric" was a key intermediary between aesthetics and ideology. As I will suggest, it

is precisely rhetoric—a figure or prosopopoeia—that is both the condition and aim of the editorial projects that I will discuss. See Warminski's "'As the Poets Do It': On the Material Sublime," in *Material Events: Paul de Man and the Afterlife of Theory* (Minneapolis: University of Minnesota Press, 2001), 3–31.

20. See Sarah Zimmerman, "Accounting for Clare," *College Literature* 62, no. 3 (January 2000): 317–34. She writes: "I am concerned with a literary historical irony: that some of the very critical practices that have served to keep Clare in view have also rendered him half visible and thereby partially inaccessible to students, scholars, and general readers. An often subtle, yet persistent emphasis on biography in Clare criticism, a convention that began as a successful marketing strategy by Clare's publisher and editor John Taylor for promoting the 'Northamptonshire peasant' (as he was identified on the title pages of his first two collections), ultimately had the less salutary effect of subordinating the poems to the appeal of a life well told" (317). Zimmerman offers an excellent account of the way that critics and editors from the beginning have either feminized or hypermasculinized Clare and in doing so both have produced Clare as a passive figure. My argument shares some of Zimmerman's interest in recognizing the "critical fiction" that is John Clare, although my focus on editing and ethics differs from hers.

21. Indeed, although Clare wrote about 3,500 poems in his lifetime, Bate's biography is nearly twice as long as his collection of poetry.

22. Bate, *John Clare*, 568.

23. Bate, *"I Am,"* xix. Curiously, as I will show later in this chapter, in the poem from which Bate draws the title to his book, the poetic subject, who we can associate with the authorial subject, does long for a position akin to childhood and registers this "return" as a utopic arrival.

24. Bate, *"I Am,"* xix.

25. Bate, *John Clare*, 565.

26. Ibid., 566.

27. R. K. R. Thornton, another of Clare's editors, also takes "raw or cooked" as the title of his review of Bate's and Robinson's selected poems. Thornton goes onto suggest that "we cannot tell always and exactly what Clare's wishes were, and we have in that case to let his texts have authority we would otherwise usurp." In other words, he suggests that we think about authority as a textual rather than an authorial effect. See R. K. R. Thornton, "Review of John Clare, *Selected Poems*," *John Clare Society Journal* 24 (2005): 82.

28. Ibid., 85.

29. Ibid.

30. Simon Kövesi, "Beyond the Language Wars," 66.

31. Ibid.

32. See Paul de Man, "Autobiography as De-Facement" in *The Rhetoric of Romanticism* (New York: Columbia University Press, 1984); and J. Hillis Miller, *Versions of Pygmalion* (Cambridge, Mass.: Harvard University Press, 1990).

33. Angus Fletcher also has reflected upon prosopopoeia in Clare although in a somewhat different context. He writes: "The chief obscurity of what happens to the factual in eighteenth-century descriptive poetry is that poets interlard facts and material

names with personifications. While this device has been much discussed, it seems we should more strongly emphasize the ancient purpose of the figure: it bridges between material and spiritual worlds, sharing in both, announcing the ideal, but always embodying some kind of personhood—the face, the masking *prosopon*, the visage of a believed ideal form. Personification seems to express our need to believe that ideas are only existentially significant if they appear to us as living beings, as thoughts capable of personal agency . . . is it possible that Clare's perceptions of the creatures of his village, Whitman's phrases of note as he confides his vision of democracy and democratic variety, Ashbery's wandering asides, while he perambulates through the landscape of his own thoughts—is it possible that these are all a new variant of ancient personifications, new *persons* always on the lookout of their own discovery?" See Angus Fletcher, *Toward a New Theory of American Poetry: Democracy, the Environment, and the Future of the Imagination* (Cambridge, Mass.: Harvard University Press, 2004), 135–36.

34. Miller, *Versions of Pygmalion*, viii, 15.

35. Bate, *John Clare*, 575.

36. Ibid., 574 (my emphasis).

37. Bate, *"I Am,"* xv.

38. Again, see Thornton and Kövesi for a discussion of the possibility or impossibility of achieving this end.

39. This absence can be understood as a quintessentially romantic predicament. See the example of the Jena Romantics, especially as discussed by Nancy and Lacoue-Labarthe and by Maurice Blanchot. Blanchot writes: "Romanticism, it is true, ends badly, but this is because it is essentially what begins and what cannot but finish badly: an end that is called suicide, madness, loss, forgetting. And certainly it is often without works, but this is because it is the work of the absence of (the) work; a poetry affirmed in the purity of the poetic act, an affirmation without duration, a freedom without realization, a force that exalts in disappearing and that is in no way discredited if it leaves no trace." Maurice Blanchot, *The Infinite Conversation,* trans. Susan Hanson (Minneapolis: University of Minnesota Press, 1993), 352–53. While this statement refers to the Jena Romantics, and can be easily related to Clare's asylum poetry (much of which was dictated), I would argue that it also can be understood to describe Clare's work more generally as "absent."

40. When de Man reads Wordsworth's *Excursion*, he notes the difference between being "deaf and mute"—i.e., an inability to hear or speak—and being "silent"—merely not speaking. One could say that the distinction between an absolute and a relative capacity to speak is parallel to the difference between an absolute absence and an historical loss. See "Autobiography as De-Facement," 80–81.

41. Dominick LaCapra, *Writing History, Writing Trauma* (Baltimore: Johns Hopkins University Press, 2001), 44.

42. Ibid., 46.

43. Ibid., 47–48, 51.

44. John Clare, *The Later Poems of John Clare, 1837–1864*, ed. Eric Robinson and David Powell (Oxford: Clarendon Press, 1984), 1:396–97. In his edition, Bate removes the double punctuation in lines 2, 3, 7; the final comma in lines 4, 7, 8; and the middle commas in lines 6, 11, 13, 16, 17. He adds a comma after "rather" in line 12 and after "above" in line 18; adds end commas in lines 13 and 14; adds an apostrophe to "lifes" in

line 10; changes the semicolon in line 15 to a comma; and replaces the possessive "s" in line 5 with a hyphen. He also spells "tost" as "tossed."

45. Timothy Morton, *Ecology without Nature: Rethinking Environmental Aesthetics* (Cambridge, Mass.: Harvard University Press, 2007), 198.

46. It is in this sense that the poem also can be understood to exemplify a certain mode of witnessing or testimony.

47. For another version of this structure, see Paul de Man's reading of Wordsworth's "A slumber did my spirit seal" in "The Rhetoric of Temporality." That both poems rely upon sleep, in Wordsworth's case a powerful vision of awakening, in Clare's case the return to sleep from suffering in "the living sea of waking dreams," may suggest that the figure of sleep as a vehicle of overcoming or restoration, insofar as it remains an overdetermined figure, has a particularly high propensity for repetition.

48. Quoted in Paul de Man, "Phenomenality and Materiality in Kant," in *Aesthetic Ideology* (Minneapolis: University of Minnesota Press, 1996), 80.

49. Ibid., 81. See also de Man, "Kant's Materialism," in *Aesthetic Ideology*, 127.

50. De Man, "Phenomenality and Materiality in Kant," 81.

51. Morton offers a similar reading of this passage, albeit in an ecologic mode. He writes: "At first glance, the closest thing we get to ecology is the last couplet where the narrator wishes for an impossible relief. And even here there is an ambiguity in the sense of 'above': is the narrator lying with the sky above him, or lying 'above . . . the sky' in heaven? But the very form of this yearning and impossibility is precisely the *most* ecological thing bout the poem" (198). In his sophisticated reading, Morton goes on to suggest that Clare's ecologism, far from determined by his attachment to the local instead is marked by an awareness of the other and by his very *displacement*. Bridget Keegan reads the end of this poem as aiming for the possibility of experiencing the "world without us" in "The World Without Us: Romanticism, Environmentalism, and Imagining Nature," in *A Companion to Romantic Poetry*, ed. Charles Mahoney (Malden, Mass.: Wiley-Blackwell, 2010), 554–68.

52. See Emile Benveniste, "Subjectivity in Language" in *Problems in General Linguistics* (Coral Gables, Fla.: University of Miami Press).

53. This is not, however, the case in comparison to the 1865 Frederick Martin edition, which is significantly different from the others:

I am! yet what I am who cares, or knows?
    My friends forsake me like a memory lost.
I am the self-consumer of my woes,
    They rise and vanish, an oblivious host,
Shadows of life, whose very soul is lost
And yet I am—I live—though I am toss'd

Into the nothingness of scorn and noise,
    Into the living sea of waking dream,
Where there is neither sense of life, nor joys,
    But the huge shipwreck of my own esteem
And all that's dear. Even those I loved the best
Are strange—nay, they are stranger than the rest.

I long for scenes where man has never trod,
For scenes where woman never smiled or wept;
There to abide with my Creator, God,
And sleep as I in childhood sweetly slept
Full of high thoughts, unborn. So let me lie,
The grass below; above the vaulted sky.

In Frederick Martin, *The Life of John Clare*, ed. Eric Robinson and Geoffrey Summerfield (London: Frank Cass & Co, 1964), 293.

4. Inventions of Self-Identity

1. John Clare, *John Clare by Himself*, ed. Eric Robinson and David Powell (Ashington, England: Mid Northumberland Arts Group, 1996), 271.

2. Agamben's thinking, though here unacknowledged, no doubt is indebted to Françoise Frontisi-Ducroux's work on ancient Greece, which he discusses in *Remnants of Auschwitz: The Witness and the Archive* (New York: Zone Books, 2000).

3. Giorgio Agamben, *Potentialities: Collected Essays in Philosophy* (Stanford: Stanford University Press, 1999), 46.

4. Ibid., 51.

5. Ibid., 53.

6. Compare with Judith Butler, *Gender Trouble: Feminism and the Subversion of Identity* (New York: Routledge, 1990).

7. Giorgio Agamben, *Potentialities*, 54.

8. Ibid.

9. Ibid., 46.

10. Clare, *John Clare by Himself*, 271.

11. Ibid.

12. Ibid., 183.

13. Unlike Cartesian or Lockean conceptions of the self that depend upon self-consciousness, Clare is anything but rational or self-identical.

14. Clare, *John Clare by Himself*, 271.

15. Burns's examples are both outrageous and understandable: How could a bridegroom forget he has just been married? How could a king forget that he is king? How could a mother forget her child? Yet, each example, for all its absurdity, is marked by its newness: The change is only an hour or a day old, and, at least in the first two cases, is the effect of a promise, decree, or oath, that is, a linguistic act that effects the difference. Yet, however justifiable or unjustifiable these acts of forgetting might be, the poem concludes with a promise of unequivocal remembrance of one whose relation to the speaker is far less obvious. What establishes the relation and the remembrance is the debt accrued over a long time.

16. Jonathan Bate, *John Clare: A Biography* (New York: Farrar, Straus & Giroux, 2003), 415.

17. Ibid. While Bridget Keegan associates the poem with the dissolution of the poetic subject, I would say that it in fact does the very opposite, for it shows the pro

found stability of a subject despite the dissolution of the world. See "Rural Poetry and the Self-Taught Tradition," in *The Blackwell Companion to Eighteenth-Century Poetry*, ed. Christine Gerrard (Oxford: Blackwell, 2006), 563.

18. This is Jonathan Bate's version of the poem. The Oxford version, which appears in *The Later Poems of John Clare, 1837–1864*, differs only by the addition in some cases and removal in other cases of punctuation and the absence of "say" in the first line.

19. Steven Miller and I take up the relation between marriage and death in our introduction to "Literature and the Right to Marriage," a special issue of *Diacritics*.

20. See, for example, Emile Benveniste, *Problems in General Linguistics* (Coral Gables, Fla.: University of Miami Press).

21. See Keegan, "Rural Poetry and the Self-Taught Tradition," for a discussion of Clare's relation to romantic transcendence.

22. João Biehl, *Vita: Life in a Zone of Social Abandonment* (Berkeley: University of California Press, 2005), 11.

23. Ibid., 24.

24. Ibid. This evokes the condition that Agamben describes in *Remnants of Auschwitz* when he defines the human as that which survives the human See my discussion of this condition in *Romanticism after Auschwitz* (Stanford: Stanford University Press, 2007).

25. See, for example, Agamben's discussion of the unnamed boy outside of language in *Remnants of Auschwitz*, Celan's poems, or Clare's asylum poems, which both do and do not belong to a language. It would seem that Catarina's dictionary, which invents a language, signals a point that is not merely that of language's unequivocal survival.

26. Biehl, *Vita*, 318.

27. Ibid., 3. The quotations that follow in this paragraph are at 318, 5, 318, 321, 350, and 317, respectively.

28. The model here is closer to Primo Levi's account of the relation between the survivor and the witness, although here there is an anthropologist or professional witness rather than one who by accident, luck, or entry into an ethical grey zone managed to survive.

29. Quoted in Bate, *John Clare*, 465.

30. Roy Porter, "'All Madness for Writing': John Clare and the Asylum," in Hugh Haughton, Adam Phillips, and Geoffrey Summerfield, *John Clare in Context* (Cambridge: Cambridge University Press, 1994), 266.

31. Ibid., 266–67.

32. Ibid., 267.

33. There is some ambivalence here, since poetry is also identified as a source.

34. Paul de Man, *Blindness and Insight: Essays in the Rhetoric of Contemporary Criticism* (Minneapolis: University of Minnesota Press, 1983), 11.

35. Ibid., 19.

36. Ibid., 18.

37. De Man's examples are "linguistics, psychoanalysis, and anthropology," which today would be considered at the humanistic edges of social science, in contrast to forms of economics, sociology, psychology, or political science, let alone literary criticism that rely extensively upon quantitative methods, including statistics and game theory. These latter disciplines focus on actors or agents, rather than subjects, whereas

the human sciences, as de Man refers to them in the essay, use narrative in a way that remains much closer to literature or the humanities. The question that remains to be taken up elsewhere is to what extent twenty-first-century social science is susceptible to the same naïveté about the status of the subject or whether its formulae and reduction of sociality to sheer number (as figure) turns out to be much closer to lyric poetry as de Man understands it. That said, the field of anthropology, to the contrary, has taken up the subject and incorporated the self into its investigations. The question endures, however, whether the anthropological approach is sufficiently literary, and indeed, whether the conception of the self at work there is an acknowledgement of the void or of power.

38. Ibid., 19.

39. For another version of this notion of criticism as essentially romantic, see Friedrich Schlegel's "Athenaeum Fragments" in *Philosophical Fragments* (Minneapolis: University of Minnesota Press, 1991) and Philippe Lacoue-Labarthe and Jean-Luc Nancy, *The Literary Absolute: The Theory of Literature in German Romanticism* (Albany: SUNY Press, 1988). See also the text of Friedrich Schelling's Earliest-System Program.

40. See Thomas Pfau, "Rhetoric and the Existential: Romantic Studies and the Question of the Subject," *Studies in Romanticism* 26, no. 4 (Winter 1987): 487–512. For Pfau, identifying the linguistic condition of the subject, however, is not an end in itself, but it exposes the relevance of romanticism (or literature) to "philosophical anthropology." As he explains: "If romanticism teaches us that subjectivity can be thought only as a linguistic and thus provisional projection by an intrinsically unstable and nonidentical consciousness, then the study of romanticism will have to relate the significance of this fact to the always fundamental question concerning man" (512).

41. In this sense, what de Man calls "philosophical anthropology" could be another name for Dasein analysis. De Man will revise this position in the later essays of the 1970s and 1980s where he dismisses forms of sublime negative knowledge such as those he finds here in Rousseau or elsewhere in Wordsworth, above all. For another approach, in the time of social science's quantitative turn, consider game theory, which relies both upon a strong fiction and a methodological approach that assumes objectivity as an effect of rigor.

42. Judith Butler, *Giving an Account of Oneself* (New York: Fordham University Press, 2005), 19.

5. The Poetics of Homelessness

1. Hugh Haughton, Adam Phillips, and Geoffrey Summerfield, *John Clare in Context* (Cambridge: Cambridge University Press, 1994), 1.

2. For another acknowledgment of the risks of transcendental accounts of homelessness, see Anthony Vidler, *The Architectural Uncanny: Essays in the Modern Unhomely* (Cambridge, Mass.: MIT Press, 1992): "faced with the intolerable state of real homelessness, any reflection on the 'transcendental' or psychological unhomely risks trivializing or, worse, patronizing social or political action" (13).

3. See the UNHCR account of "internally displaced people": "internally displaced people, or IDPs, are often wrongly called refugees. Unlike refugees, IDPs have not crossed an international border to find sanctuary, but have remained inside their home

countries. . . . As citizens, they retain all of their rights and protection under both human rights and international humanitarian law." www.unhcr.org/pages/49c3646c146.html. Accessed January 23, 2015.

4. For a resonant account that perceives romanticism in terms of posterity, see Andrew Bennett, *Romantic Poets and the Culture of Posterity* (Cambridge: Cambridge University Press, 1999). For a discussion of the long history of making Clare heard, one that offers some insight into the rhetorical frame of Phillips and Haughton's introduction, see Chapter 3.

5. Douglas Chambers, "'A love for every simple weed': Clare, Botany, and the Poetic Language of Lost Eden," in Haughton, Phillips, and Summerfield, *John Clare in Context*, 238.

6. Jonathan Bate, *John Clare: A Biography* (New York: Farrar, Straus & Giroux, 2003), 387.

7. On "uncounted experience," see Anne-Lise François, *Open Secrets: The Literature of Uncounted Experience* (Stanford: Stanford University Press, 2007), and Brian M. McGrath, *The Poetics of Unremembered Acts: Reading, Lyric, Pedagogy* (Evanston, Ill.: Northwestern University Press, 2013).

8. Rob Nixon, *Slow Violence and the Environmentalism of the Poor* (Cambridge, Mass.: Harvard University Press, 2011), 19.

9. Ibid.

10. John Barrell, *The Idea of Landscape and the Sense of Place, 1730–1840: An Approach to the Poetry of John Clare* (Cambridge: Cambridge University Press, 1972), 215.

11. Ibid.

12. Ibid., 295.

13. Nixon, *Slow Violence and the Environmentalism of the Poor*, 19; my emphasis.

14. John Clare, *By Himself*, ed. Eric Robinson and David Powell (Ashington, England: Mid Northumberland Arts Group, 1996), 40.

15. For an example of this poetry, see "The Moors."

16. Clare, *By Himself*, 40–41; my emphasis.

17. Bloomfield, however, does not speak in his own voice at this moment, but cites Thomas Paine. See Barrell, *The Idea of Landscape and the Sense of Place*, 120–21.

18. Ibid., 121.

19. See, for comparison, the way in which Derrida writes about the "event" as an event that cannot be known apart from retroactively.

20. Bate, *John Clare*, 42.

21. For maps, see ibid., 40; Clare, *By Himself*, xxii–iii; Barrell, *The Idea of Landscape and the Sense of Place*, 102 and 107; reprinted in Franco Moretti, *Graphs, Maps, Trees: Abstract Models for a Literary History* (London: Verso, 2007), 40–41.

22. Clare, *By Himself*, 41.

23. Sigmund Freud, "The Uncanny," *The Standard Edition of the Complete Psychological Works of Sigmund Freud* (London: Hogarth Press and the Institute of Psycho-analysis, 1953), 17:219 n. 1. Terry Castle has argued for an earlier, Enlightenment origin of the uncanny. See also the discussion in Nicholas Royle, *The Uncanny: An Introduction* (New York: Manchester University Press, 2003).

24. Freud, "The Uncanny," 224.

25. See Ross Wilson, "Clare's Indistinct Array," *Romanticism* 17 (2011): 148–59.

26. We could assign names to this rethinking: Freud, Heidegger, Arendt. These names and others—Hölderlin, Hartman, Agamben, Adorno, Derrida, Szendy—and the bodies of work that they evoke lurk in the shadows of my reading of Clare, not only informing it but also read through it.

27. Chambers, "'A love for every simple weed,'" 238.

28. See Barrell's discussion of this, referring to the poem "Remembrances": "There's something of the same confusion about this poem that we noticed in the early poems about the enclosure in which it's hard to grasp quite what Clare's nostalgia is for: whether it is for a vision of Helpston that has inevitably left him as he has grown older, or for a Helpston which has undergone the concrete change of being enclosed" (*The Idea of Landscape and the Sense of Place*, 175). Barrell goes on to describe enclosure as an emblem for loss, which is not how I understand the confusion. Rather, it is closer to what I describe in *Romanticism after Auschwitz* as a structure of survival, one that cannot be contained by the conceptual or poetic frameworks available for its recollection.

29. In the published edition of *The Rural Muse*, only eight of the poem's twenty seven stanzas appear, and the poem bears the date June 20, 1832.

30. John Clare, *The Rural Muse*, ed. R. K. R. Thornton (Manchester: Carcanet New Press, 1982), 147.

31. The *OED* defines this usage of *of* as "denoting a person or a thing that is distinguished out of a number, or out of all, on account of excellence . . . as in the Hebraistic *Song of songs*." Curiously, in the section of the poem that appeared in the printed edition of *The Rural Muse*, Clare only uses "of" three times—in the title, in the first line, and as the first word of the fifth stanza, "Of happiness, and thoughts arise / With home-bred pictures many a one." Yet throughout, Clare uses what Barbara Strang calls the "-*s*-genitive" in a manner that his initial readers found "prominent (and objectionable)," because he used them "with inanimate nouns even abstracts, which normally prefer the *of*- construction"; see her essay "John Clare's Language" in *The Rural Muse*, 171. Thus, in this first stanza, he writes "hazel's happy green"; "bluebell's quiet-hanging blooms"; and "envy's sneer." Each of these examples is tied to a personification of the *terroir*.

32. Etymologically, "wane," which also describes the rhythms of the moon, is, like "vanity," derived from the Latin *vanus*, a connection that at the level of the word reconnects poetry (here on the wane) with life ("vanity of vanities").

33. Jonathan Bates, *"I Am": The Selected Poetry of John Clare* (New York: Farrar, Straus & Giroux, 2003), 180.

34. Ibid., 181.

35. Ibid.

36. Ibid., 182.

37. Clare, *The Rural Muse*, 152.

38. For one among the many efforts to think and rethink cosmopolitanism's relation to universalism, see Peter Szendy, *Kant in the Land of Extraterrestrials: Cosmopolitical Philosofictions* (New York: Fordham University Press, 2013).

39. Clare, *The Rural Muse*, 152.

40. James Griffioen, "Feral Houses." *Sweet Juniper*. July 23, 2009. Web. Accessed January 28, 2015.

41. James Griffioen, "(More) Feral Houses." *Sweet Juniper.* June 18, 2010. Web. Accessed January 28, 2015.

42. Clare, *The Rural Muse*, 152.

43. Hannah Arendt, *The Origins of Totalitarianism* (New York: Harcourt Brace Jovanovich, 1966), 298–99; Giorgio Agamben, *Homo Sacer: Sovereign Power and Bare Life*, trans. Daniel Heller-Roazen (Stanford: Stanford University Press, 1998), 134.

44. Giorgio Agamben, "We Refugees," *Symposium* 49, no. 2 (Summer 1995): 114–19.

Bibliography

Adorno, Theodor W. *Minima Moralia: Reflections from Damaged Life.* Translated by E. F. N. Jephcott. London: Verso, 1978.

Agamben, Giorgio. *The Coming Community.* Translated by Michael Hardt. Minneapolis: University of Minnesota Press, 1993.

———. *Homo Sacer: Sovereign Power and Bare Life.* Translated by Daniel Heller-Roazen. Stanford: Stanford University Press, 1998.

———. *Infancy and History: The Destruction of Experience.* Translated by Liz Heron. New York: Verso, 1993.

———. *Means Without End: Notes on Politics.* Translated by Vincenzo Binetti and Cesare Casarino. Minneapolis: University of Minnesota Press, 2000.

———. *Potentialities: Collected Essays in Philosophy.* Translated by Daniel Heller-Roazen. Stanford: Stanford University Press, 1999.

———. *Remnants of Auschwitz: The Witness and the Archive.* Translated by Daniel Heller-Roazen. New York: Zone Books, 2000.

———. "We Refugees." Translated by Michael Rocke. *Symposium* 49, no. 2 (Summer 1995): 114–119.

Arendt, Hannah. *The Origins of Totalitarianism.* New York: Harcourt Brace Jovanovich, 1966.

Ashbery, John. *Other Traditions.* Cambridge, Mass.: Harvard University Press, 2000.

Bajorek, Jennifer. *Counterfeit Capital: Poetic Labor and Revolutionary Irony.* Stanford: Stanford University Press, 2009.

Balibar, Etienne. "'Possessive Individualism' Reversed: From Locke to Derrida." *Constellations* 9, no. 3 (2002): 299–317.

———. *We, the People of Europe? Reflections on Transnational Citizenship.* Translated by James Swenson. Princeton, N.J.: Princeton University Press, 2003.

Ball, Joanna. "'The Tear Drops on the Book I Read': John Clare's Reading in the Northampton General Lunatic Asylum, 1841–1864." *Wordsworth Circle* 34, no. 3 (Summer 2003): 155–58.

Barrell, John. *The Idea of Landscape and the Sense of Place, 1730–1840: An Approach to the Poetry of John Clare.* Cambridge: Cambridge University Press, 1972.

Bate, Jonathan. *"I Am": The Selected Poetry of John Clare.* New York: Farrar, Straus & Giroux, 2003.

———. *John Clare: A Biography.* New York: Farrar, Straus & Giroux, 2003.

Bennett, Andrew. *Romantic Poets and the Culture of Posterity.* Cambridge: Cambridge University Press, 1999.

Benveniste, Emile. *Problems in General Linguistics.* Coral Gables, Fla.: University of Miami Press, 1973.

Berlant, Lauren. *Cruel Optimism.* Durham, N.C.: Duke University Press, 2011.

———. *The Queen of America Goes to Washington City: Essays on Sex and Citizenship.* Durham, N.C.: Duke University Press, 1997.

Biehl, João. *Vita: Life in a Zone of Social Abandonment.* Berkeley: University of California Press, 2005.

Blanchot, Maurice. *The Infinite Conversation.* Translated by Susan Hanson. Minneapolis: University of Minnesota Press, 1992.

———. *The Instant of My Death.* Translated by Elizabeth Rottenberg. Stanford: Stanford University Press, 2000.

Brooks, Peter. *Enigmas of Identity.* Princeton, N.J.: Princeton University Press, 2011.

Burwick, Frederick. *Poetic Madness and the Romantic Imagination.* University Park: Pennsylvania State University Press, 1996.

Butler, Judith. *Gender Trouble: Feminism and the Subversion of Identity.* New York: Routledge, 1990.

———. *Giving an Account of Oneself.* New York: Fordham University Press, 2005.

Cameron, Sharon. *Impersonality: Seven Essays.* Chicago: University of Chicago Press, 2007.

Caruth, Cathy. *Empirical Truths and Critical Fictions: Locke, Wordsworth, Kant, Freud.* Baltimore: Johns Hopkins University Press, 1991.

Castle, Terry. *The Female Thermometer: Eighteenth-Century Culture and the Invention of the Uncanny.* New York: Oxford University Press, 1995.

Chambers, Douglas. "'A love for every simple weed': Clare, Botany, and the Poetic Language of Lost Eden." In Haughton, Phillips, and Summerfield, *John Clare in Context,* 238–58.

Chase, Cynthia. *Romanticism.* New York: Longman, 1993.

Cheah, Pheng, and Bruce Robbins, eds. *Cosmopolitics: Thinking and Feeling Beyond the Nation.* Minneapolis: University of Minnesota Press, 1998.

Cherry, J. L. *Life and Remains of John Clare, "The Northamptonshire Poet."* Northampton: J. Taylor & Son, 1873.

Chilcott, Tim. *"A Real World & Doubting Mind": A Critical Study of the Poetry of John Clare.* Hull: Hull University Press, 1985.

Chirico, Paul. *John Clare and the Imagination of the Reader.* New York: Palgrave Macmillan, 2007.

Clare, Johanne. *John Clare and the Bounds of Circumstance.* Kingston, Ontario: McGill-Queen's University Press, 1987.

Clare, John. *Autobiographical Writings.* Edited by Eric Robinson. New York: Oxford University Press, 1983.

———. *By Himself.* Edited by Eric Robinson and David Powell. Ashington, England: Mid Northumberland Arts Group, 1996.

———. *A Champion for the Poor: Political Verse and Prose.* Edited by Eric Robinson, P. M. S. Dawson, and David Powell. Manchester: Carcanet, 2000.

———. *The Early Poems of John Clare, 1804–1822.* Edited by Eric Robinson, David Powell, and Margaret Grainger. Oxford: Oxford University Press, 1989.

———. *The Journal; Essays; The Journey from Essex.* Manchester: Carcanet New Press, 1980.

———. *The Later Poems of John Clare, 1837–1864.* Edited by Eric Robinson and David Powell. New York: Oxford University Press, 1984.

———. *Major Works*. Edited by Eric Robinson and David Powell. Oxford: Oxford University Press, 2008.

———. *The Midsummer Cushion*. Ashington: Carcanet Press, 1979.

———. *Northborough Sonnets*. Edited by Eric, Robinson, P. M. S. Dawson, and David Powell. Manchester: Carcanet, 1995.

———. *Poems of the Middle Period, 1822–1837*. Edited by Eric Robinson, P. M. S. Dawson, and David Powell. New York: Oxford University Press, 1996.

———. *The Rural Muse: Poems*. Edited by R. K. R. Thornton. Manchester: Carcanet Press, 1982.

———. *The Shepherd's Calendar*. New York: Oxford University Press, 1964.

Clark, Timothy. *The Theory of Inspiration*. Manchester: Manchester University Press, 1997.

Cohen, Tom, et al. *Material Events: Paul de Man and the Afterlife of Theory*. Minneapolis: University of Minnesota Press, 2001.

Cooke, Brett, and Frederick Turner. *Biopoetics: Evolutionary Explorations in the Arts*. Lexington, Ky.: Paragon House, 1999.

Culler, Jonathan. *The Pursuit of Signs: Semiotics, Literature, Deconstruction*. Ithaca, N.Y.: Cornell University Press, 2002.

Deleuze, Gilles. *Essays Critical and Clinical*. Translated by Daniel W. Smith and Michael A. Greco. Minneapolis: University of Minnesota Press, 1997.

De Man, Paul. *Aesthetic Ideology*. Minneapolis: University of Minnesota Press, 1996.

———. *Blindness and Insight: Essays in the Rhetoric of Contemporary Criticism*. Minneapolis: University of Minnesota Press, 1983.

———. *The Resistance to Theory*. Minneapolis: University of Minnesota Press, 1986.

———. *The Rhetoric of Romanticism*. New York: Columbia University Press, 1984.

De Man, Paul, Kevin Newmark, Andrzej Warminski, and E. S. Burt. *Romanticism and Contemporary Criticism: The Gauss Seminar and Other Papers*. Baltimore: Johns Hopkins University Press, 1996.

Derrida, Jacques. *The Beast and Sovereign*. Translated by Geoffrey Bennington. Chicago: University of Chicago Press, 2009–11.

———. *Monolingualism of the Other, or The Prosthesis of Origin*. Translated by Patrick Mensah. Stanford: Stanford University Press, 1998.

Deutscher, Penelope. "The Inversion of Exceptionality: Foucault, Agamben, and 'Reproductive Rights.'" *South Atlantic Quarterly* 107, no. 1 (2008): 55–70.

Edelman, Lee. *No Future: Queer Theory and the Death Drive*. Durham, N.C.: Duke University Press, 2004.

Elfenbein, Andrew. *Romanticism and the Rise of English*. Stanford: Stanford University Press, 2009.

Esposito, Roberto. *Bíos: Biopolitics and Philosophy*. Translated by Timothy Campbell. Minneapolis: University of Minnesota Press, 2008.

Fletcher, Angus. *A New Theory for American Poetry: Democracy, the Environment, and the Future of Imagination*. Cambridge, Mass.: Harvard University Press, 2004.

Foucault, Michel. *History of Madness*. Translated by Jonathan Murphy and Jean Khalfa. London: Routledge, 2006.

———. *The History of Sexuality*. Translated by Robert Hurley. New York: Vintage Books, 1990.

———. *"Il Faut Défendre La Société": Cours Au Collège de France (1975–1976).* Paris: Seuil/Gallimard, 1997.

———. *"Society Must Be Defended": Lectures at the Collège de France, 1975–1976.* New York: Picador, 2003.

Foulds, Adam. *The Quickening Maze.* London: Jonathan Cape, 2009.

François, Anne-Lise. *Open Secrets: The Literature of Uncounted Experience.* Stanford: Stanford University Press, 2007.

Freud, Sigmund. *The Standard Edition of the Complete Psychological Works of Sigmund Freud.* London: Hogarth Press and the Institute of Psycho-analysis, 1953.

Galton, Francis. *Hereditary Genius; an Inquiry into Its Laws and Consequences.* Cleveland: World Publishing, 1962.

Gigante, Denise. *Life: Organic Form and Romanticism.* New Haven: Yale University Press, 2009.

Gilroy, Paul. *Postcolonial Melancholia.* New York: Columbia University Press, 2005.

Goodridge, John. *John Clare and Community.* Cambridge: Cambridge University Press, 2013.

Goodridge, John, ed. *The Independent Spirit: John Clare and the Self-Taught Tradition.* Helpston, England: John Clare Society and the Margaret Grainger Memorial Trust, 1994.

Goodridge, John, and Simon Kövesi, eds. *John Clare: New Approaches.* Peterborough, England: John Clare Society, 2000.

Gorji, Mina. "Clare's Awkwardness." *Essays in Criticism: A Quarterly Journal of Literary Criticism* 54, no. 3 (2004): 216–39.

———. *John Clare and the Place of Poetry.* Liverpool: Liverpool University Press, 2008.

Gorji, Mina, and Kirstie Blair. *Class and the Canon: Constructing Labouring-Class Poetry and Poetics, 1780–1900.* New York: Palgrave Macmillan, 2013.

Grainger, Mararet. *The Natural History Prose Writings of John Clare.* Oxford: Clarendon Press, 1983.

Greaney, Patrick. *Untimely Beggar: Poverty and Power from Baudelaire to Benjamin.* Minneapolis: University of Minnesota Press, 2008.

Griffioen, James. *Sweet Juniper.* Web. Accessed January 28, 2015.

Grigson, Geoffrey. *Poems of John Clare's Madness.* London: Routledge and Kegan Paul, 1949.

Guyer, Sara. *Romanticism after Auschwitz.* Stanford: Stanford University Press, 2007.

Hacking, Ian. *Mad Travelers: Reflections on the Reality of Transient Mental Illness.* Cambridge, Mass.: Harvard University Press, 1998.

Hartman, Geoffrey H. *The Unremarkable Wordsworth.* Minneapolis: University of Minnesota Press, 1987.

Haughton, Hugh, Geoffrey Summerfield, and Adam Phillips. *John Clare in Context.* Cambridge: Cambridge University Press, 1994.

Helsinger, Elizabeth. "Clare and the Place of the Peasant Poet." *Critical Inquiry* 13, no. 3 (1987): 509–31.

Heyes, Bob. "Writing Clare's Poems: 'The Myth of Solitary Genius.'" In Goodrich and Kövesi, *John Clare: New Approaches*, 33–46.

Jackson, John Brinckerhoff. *Discovering the Vernacular Landscape.* New Haven: Yale University Press, 1984.

Jacobs, Carol. *Uncontainable Romanticism: Shelley, Brontë, Kleist*. Baltimore: Johns Hopkins University Press, 1989.

Jin, Ha. *The Writer as Migrant*. Chicago: University of Chicago Press, 2008.

Johnson, Barbara. *Persons and Things*. Cambridge, Mass.: Harvard University Press, 2008.

———. *A World of Difference*. Baltimore: Johns Hopkins University Press, 1987.

Jöttkandt, Sigi. *First Love: A Phenomenology of the One*. Melbourne: RE:Press, 2010.

Keegan, Bridget. "Rural Poetry and the Self-Taught Tradition." In *The Blackwell Companion to Eighteenth-Century Poetry*, edited by Christine Gerrard. Oxford: Blackwell, 2006, 563–76.

———. "The World without Us: Romanticism, Environmentalism, and Imagining Nature." In Mahoney, 554–71. Chichester, England: Wiley-Blackwell, 2011.

———. "Boys, Marvellous Boys: John Clare's 'Natural Genius.'" In Goodrich and Kövesi, *John Clare: New Approaches*, 65–76.

Keller, Evelyn Fox. *The Mirage of a Space between Nature and Nurture*. Durham, N.C.: Duke University Press, 2010.

Kelley, Theresa M. *Clandestine Marriage: Botany and Romantic Culture*. Baltimore: Johns Hopkins University Press, 2012.

———. "Postmodernism, Romanticism, and John Clare." In *The Lessons of Romanticism: A Critical Companion*, edited by Thomas Pfau and Robert F. Gleckner. Durham, N.C.: Duke University Press, 1998, 157–70.

Kornbluh, Anna. *Realizing Capital: Financial and Psychic Economies in Victorian Form*. New York: Fordham University Press, 2014.

Kövesi, Simon. "Beyond the Language Wars: Towards a Green Edition of John Clare." *John Clare Society Journal* 26, no. 26 (2007): 61–75.

———. "The John Clare Copyright: 1820–2000." *Wordsworth Circle* 31, no. 3 (2000): 112–19.

LaCapra, Dominick. *History in Transit: Experience, Identity, Critical Theory*. Ithaca, N.Y.: Cornell University Press, 2004.

Lacoue-Labarthe, Philippe, and Jean-Luc Nancy. *The Literary Absolute: The Theory of Literature in German Romanticism*. Albany: SUNY Press, 1988.

Levi, Primo. *The Drowned and the Saved*. Translated by Stuart Woolf. New York: Vintage International, 1989.

Mahoney, Charles. *A Companion to Romantic Poetry*. Chichester, England: Wiley-Blackwell, 2010.

Mahood, M. M. "John Clare: The Poet as Raptor." *Essays in Criticism: A Quarterly Journal of Literary Criticism* 48, no. 3 (1998): 201–23.

Malabou, Catherine. *Plasticity at the Dusk of Writing Dialectic, Destruction, Deconstruction*. Translated by Carolyn Shread. New York: Columbia University Press, 2010.

———. *What Should We Do with Our Brain?* New York: Fordham University Press, 2008.

Martin, Frederick. *The Life of John Clare*. London: F. Cass, 1964.

McGrath, Brian M. *The Poetics of Unremembered Acts: Reading, Lyric, Pedagogy*. Evanston, Ill.: Northwestern University Press, 2013.

McKusick, James C. "John Clare and the Tyranny of Grammar." *Studies in Romanticism* 33, no. 2 (1994): 255–77.

———. "'A Language That Is Ever Green': The Ecological Vision of John Clare."

University of Toronto Quarterly: A Canadian Journal of the Humanities 61, no. 2 (1991): 226–49.

McLane, Maureen N. *Balladeering, Minstrelsy, and the Making of British Romantic Poetry*. Cambridge: Cambridge University Press, 2008.

Miller, Eric. "Enclosure and Taxonomy in John Clare." *SEL: Studies in English Literature, 1500–1900* 40, no. 4 (2000): 635–57.

Miller, J. Hillis. *The Ethics of Reading: Kant, De Man, Eliot, Trollope, James, and Benjamin*. New York: Columbia University Press, 1987.

———. *Versions of Pygmalion*. Cambridge, Mass.: Harvard University Press, 1990.

Mills, Catherine. *Futures of Reproduction*. Dordrecht: Springer, 2011.

Moretti, Franco. *Graphs, Maps, Trees: Abstract Models for a Literary History*. London: Verso, 2007.

Morton, Timothy. *Ecology without Nature: Rethinking Environmental Aesthetics*. Cambridge, Mass.: Harvard University Press, 2007.

Nisbet, John Ferguson. *The Insanity of Genius*. London: Grant and Richards, 1900.

Nixon, Rob. *Slow Violence and the Environmentalism of the Poor*. Cambridge, Mass.: Harvard University Press, 2011.

Nord, Deborah. *Gypsies and the British Imagination, 1807–1930*. New York: Columbia University Press, 2006.

Paulin, Tom. *Minotaur: Poetry and the Nation State*. Cambridge, Mass.: Harvard University Press, 1992.

Pfau, Thomas. "Reply to Theresa Kelley's 'Romantic Interiority and Cultural Objects' Romantic Circles Praxis Series." In *Romanticism and Philosophy in an Historical Age*, edited by Karen Weisman. College Park: University of Maryland, 1999.

———. "Rhetoric and the Existential: Romantic Studies and the Question of the Subject." *Studies in Romanticism* 26, no. 4 (Winter 1987): 487–512.

Porter, Roy. "'All madness for writing': John Clare and the Asylum." in Haughton, Phillips, and Summerfield, *John Clare in Context*, 259–78.

Rancière, Jacques. *The Philosopher and His Poor*. Durham, N.C.: Duke University Press, 2007.

———. *Staging the People: The Proletarian and His Double*. London: Verso, 2011.

Redfield, Marc, ed. *Legacies of Paul de Man*. New York: Fordham University Press, 2007.

Redfield, Marc. *The Politics of Aesthetics: Nationalism, Gender, Romanticism*. Stanford: Stanford University Press, 2003.

———. *The Rhetoric of Terror: Reflections on 9/11 and the War on Terror*. New York: Fordham University Press, 2009.

Rigby, Catherine E. *Topographies of the Sacred: The Poetics of Place in European Romanticism*. Charlottesville: University of Virginia Press, 2004.

Ringler, William. "Poeta Nascitur Non Fit: Some Notes on the History of an Aphorism." *Journal of the History of Ideas* 2, no. 4 (October 1941): 497–504.

Robinson, Eric. "Editorial Problems in John Clare." *John Clare Society Journal* 2 (1983): 9–23.

Robinson, Eric, and Geoffrey Summerfield. "John Taylor's Editing of Clare's *The Shepherd's Calendar*." *The Review of English Studies* 14, no. 56 (November 1963): 359–69.

Royle, Nicholas. *The Uncanny: An Introduction*. New York: Manchester University Press, 2003.

Santner, Eric L. *The Royal Remains: The People's Two Bodies and the Endgames of Sovereignty*. Chicago: University of Chicago Press, 2011.

Schiller, Friedrich. *On the Aesthetic Education of Man, in a Series of Letters*. Translated by L.A. Willoughby, and Elizabeth M. Wilkinson. Oxford: Clarendon Press, 1967.

Schlegel, Friedrich. *Philosophical Fragments*. Translated by Peter Firchow. Minneapolis: University of Minnesota Press, 1991.

Shelley, Mary Wollstonecraft. *Frankenstein, or, the Modern Prometheus: The 1818 Text*. New York: Oxford University Press, 1998.

Sinclair, Iain. *Edge of the Orison: In the Traces of John Clare's "Journey out of Essex."* London: Penguin, 2005.

Storey, Mark. *Clare: The Critical Heritage*. London and Boston: Routledge & Kegan Paul, 1973.

———. *John Clare, Selected Letters*. New York: Oxford University Press, 1988.

———. *The Letters of John Clare*. New York: Oxford University Press, 1985.

Sychrava, Juliet. *Schiller to Derrida: Idealism in Aesthetics*. Cambridge: Cambridge University Press, 1989.

Symons, Arthur, ed. *Poems by John Clare*. London: H. Frowde, 1908.

Szendy, Peter. *Kant in the Land of Extraterrestrials: Cosmopolitical Philosofictions*. New York: Fordham University Press, 2013.

Thornton, R. K. R. "Review of John Clare, *Selected Poems*." *John Clare Society Journal* 24 (2005): 82.

Tibble, John William, and Anne Northgrave Tibble. *John Clare: A Life*. Totowa, N.J.: Rowman & Littlefield, 1972.

Vardy, Alan D. *John Clare, Politics and Poetry*. Basingstoke, England: Palgrave Macmillan, 2003.

Vidler, Anthony. *The Architectural Uncanny: Essays in the Modern Unhomely*. Cambridge, Mass.: MIT Press, 1992.

Williams, Raymond. *The Country and the City*. London: Chatto and Windus, 1973.

Wilson, Ross. "Clare's Indistinct Array." *Romanticism* 17 (2011): 148–59.

Wordsworth, William, James Butler, and Karen Green. *Lyrical Ballads, and Other Poems, 1797–1800*. Ithaca, N.Y.: Cornell University Press, 1993.

Wordsworth, William, Stephen Gill, M. H. Abrams, and Jonathan Wordsworth. *The Prelude, 1799, 1805, 1850: Authoritative Texts, Context and Reception, Recent Critical Essays*. New York: Norton, 1979.

Zimmerman, Sarah M. "Accounting for Clare." *College English* 62, no. 3 (2000): 317–34.

———. *Romanticism, Lyricism, and History*. Albany: SUNY Press, 1999.

Index

Z LIT

Sara Guyer and Brian McGrath, series editors

Sara Guyer, *Reading with John Clare: Biopoetics, Sovereignty, Romanticism*

Philippe Lacoue-Labarthe, *Ending and Unending Agony: On Maurice Blanchot*. Translated by Hannes Opelz.